The Origin of the Planck Length, Planck Mass and Planck Time

A New Candidate For Dark Matter

By

Shelton W. Riggs, Jr.

The Origin of the Planck Length, Planck Mass and Planck Time

A New Candidate for Dark Matter

ISBN-13: 978-1-44954-374-7

ISBN-10: 1-44954-374-X

This book is dedicated to
Physicists, Astronomers and
Cosmologists who are curious
about the possible origins
of dark matter

About The Author

Shelton W. Riggs, Jr. earned undergraduate (University of Texas) and graduate (Vanderbilt) degrees in both Physics and Mathematics.

Professionally, he has consulted as both a hardware and software design engineer to numerous Fortune 500 companies for a wide range of scientific applications. He helped solve several scientific problems for US Army, Air Force and Navy.

Other interests include theoretical physics including quantum mechanics, relativistic mechanics and theoretical mathematics (especially the mystery of prime numbers).

Hobbies include dancing, karaoke, juggling, playing keyboards, writing songs, and writing poetry.

Other Works By Author

The Scientific Theory of God – A bridge Between Faith and Physics provides the reader with basic scientific understanding, interpretation, clarification and answers about concepts and beliefs associated with a Supreme Being. These ideas are developed and based on current theory and the standard model of physics. This new basis has revealed surprising relationships between the scientific definitions of both God and man. A model for the behavior of living matter (bioenergy) has been extended to include the behavior of human beings in terms of perception, decision and action. These concepts combined with the operation of short and long-term memory explain both human consciousness and how the mind controls the body. This model also includes how any desired behavior (provided it does not go against survival) may be achieved. This book offers a scientific creation theory and shows how it is compatible with both the big bang as well as evolutionary theory.

An Alternate Lorentz Invariant Relativistic Wave Equation offers an invariant form which differs from both the Dirac equation as well as the Klein-Gordon equation. Unlike, Schrödinger's non-relativistic wave equation, both the Dirac as well as the Klein-Gordon equation predict wave functions which do not collapse when applied to free systems at rest. On the other hand, Schrödinger's equation predicts wave functions that do collapse when applied to free systems at rest. The alternate relativistic wave equation offered by the author follows Schrödinger's philosophy that if a free system is at rest, then it is a particle with a collapsed wave function.

Nature of the First Cause – The Discovery of What Triggered the Big Bang contains the formal scientific theory of how the universe got started. It lays down the mathematical foundation for the creation theory put forth in "The Scientific Theory of God" book. It resolves the asymmetry problem of physics. It solves the two main cosmological problems by identifying both dark energy and dark matter. This theory predicts the correct order of magnitude for the number of galaxies and stars in the universe revealed by the Hubble ultra deep field results. It uncovers two entangled parallel worlds consisting of negative antimatter and positive matter. It explains the accelerated expansion of both matter and negative antimatter. It predicts the distance between matter and negative antimatter to be the Schwarzschild diameter of the expanding universe.

Primal Proofs offer several proofs that deal with prime numbers. A proof by contradiction of Goldbach's binary conjecture that every even natural number greater than two (2) can be expressed as the sum of two (2) primes is given. A proof of Goldbach's ternary conjecture that all natural numbers greater than five (5) are the sum of three (3) primes via the binary proof is presented. A proof by construction (utilizing the proof of the binary conjecture) of the twin prime conjecture is offered. A proof of the Riemann hypothesis by deduction is presented. A proof that any prime greater than three (3) is the mean of two other primes is presented. A proof is offered that any even number greater than twelve (12) satisfies Goldbach's binary conjecture in a plurality of ways. Two entangled formulas that generate all the primes beyond the second prime ($P_2 = 3$) are developed and summarized.

Acknowledgments

I acknowledge God for providing all the resources necessary to describe and predict how photons can get trapped in each others gravitational field.

I acknowledge my country for providing me with my freedoms, especially my freedom of speech.

I acknowledge my parents for providing a secure and nurturing environment that initially made my learning fun.

I acknowledge all my teachers especially my science and mathematics teachers.

I acknowledge every author in the reference section of this book for providing both ideas and data.

Preface

This study was undertaken to investigate whether or not it is possible for two identical photons to gravitationally interact with each other. It is experimentally known that a photon will interact gravitationally with a massive body such as the sun. If a photon can exchange gravitons with massive bodies, it should be possible for two photons to exchange gravitons with each other. In turn, this possibility gives rise to new candidates for dark matter.

The author has made minimal assumptions for the case of a photon moving under the influence of another photon's gravitational field.

It is known that a photon has no rest mass. However, a photon in motion does have mass according to its energy. Its mass can be calculated by dividing its energy by the speed of light squared. Its energy is proportional to its angular frequency with the constant of proportionality being Planck's constant divided by 2π.

If both photons have mass while in flight, then the gravitational force is proportional to the product of their masses and inversely proportional to the square of the distance between them.

This presentation uses both classical point mechanics as well as a relativistic quantum mechanical approach.

Solutions of a system composed of two identical photons which are trapped in each other's gravitational field are developed. The solution applies to any pair of identical particles having zero rest mass. One solution was found by treating two photons as point particles. The quantum mechanical solution came about by treating the two photons as waves. In both solutions, the predicted distance between photons was found to be proportional to the Planck length. The period of the photon's orbit was proportional to the Planck time and the mass energy of each photon is proportional to the Planck mass. Thus, the concept of a Planck length, Planck mass and Planck time all emerge from this single model.

The fundamental physical laws, basic units, physical constants and basic elementary particles have been included for reference after the glossary.

Please refer to the glossary for the definitions, values and symbols for the various physical quantities within this manuscript.

Table of Contents

Chapter 1
Point Particle System

Point Mechanical Preliminaries

For an orbiting two body system of mass m_1 and m_2 specified by two points at r_1 and r_2 respectively, the centrifugal force, F is equivalent to the force of attraction toward the center of curvature. Referring to figure 1, in the center of mass (CM) system we can generally write

(1.0) $m_1v_1^2/r_1 = F = m_2v_2^2/r_2$ and

(1.01) $v_1/r_1 = v_2/r_2 = \omega$

where ω is the angular velocity of both m_1 and m_2.

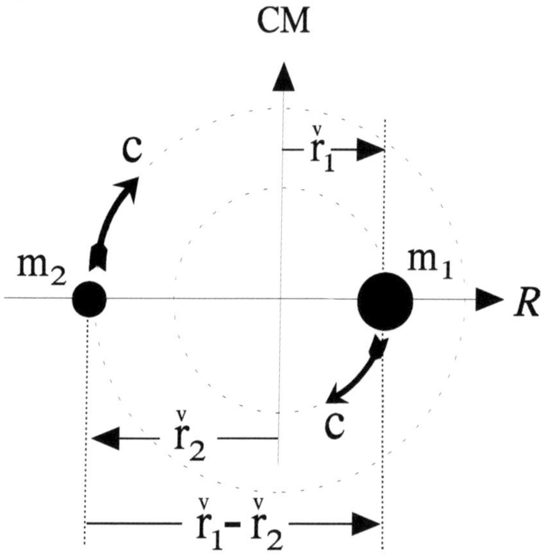

Figure 1 Center of mass (CM) of two particles influenced by each other's gravitational field.

Center of Mass (CM) System

Figure 1 shows both particles m_1 and m_2 with $m_1 > m_2$. Each particle orbits the center of mass of both systems and moves with respective velocities of v_1 and v_2. Note that $v_2 > v_1$ since $m_1 > m_2$.

When the vector position of particle 1 is at $\check{r}_1$, then simultaneously, in the center of mass (CM) system, the vector position of particle 2 is $\check{r}_2$. The curved arrows show the direction of both particles. The vertical axis (**CM**) denotes the center of mass of both particles and where it crosses the horizontal axis, is the point which represents the mass of both particles. The horizontal axis, ***R*** passes through the centers of both particles.

System Dynamical Specification

Let F be the attractive force of gravity. This means that

$$(1.1)\ F = Gm_1m_2/r_{12}^2$$

where r_{12} is the distance between m_1 and m_2. The gravitational potential energy between m_1 and m_2 is

$$(1.1.1)\ V_{12} = -\ Gm_1m_2/r_{12}$$

Equating equation (1.0) with (1.1) yields

(1.2) $m_1v_1^2/r_1 = Gm_1m_2/r_{12}^2 = m_2v_2^2/r_2$

To be relativistically correct, both masses are functions of their velocity by

(1.3) $m = m_0[1 - (v/c)^2]^{-1/2}$

where m is its mass in motion, m_0 is its mass while at rest, v is its velocity and c is the speed of light in vacuum.

Conservation of energy also requires that

(1.4) $E_1 + E_2 + V_{12} = \text{Constant} = K$

and can be explicitly expressed as

(1.5) $m_1c^2 + m_2c^2 - Gm_1m_2/r_{12} = K$

Equations

(1.2) $m_1v_1^2/r_1 = Gm_1m_2/r_{12}^2 = m_2v_2^2/r_2$

and

$$(1.5)\ m_1c^2 + m_2c^2 - Gm_1m_2/r_{12} = K$$

represent the dynamical specification of the point relativistic mechanical system of two masses (m_1 & m_2) separated by a distance of r_{12} and orbiting one another under the influence of each other's gravitational field. Thus, figure 1 is the general representation of a system of two point particles trapped in each other's gravitational field.

Chapter 2
Point Particle Solution

Photon Model Description

Consider 2 photons moving under the influence of each other's gravitational field. Each photon will orbit about the center of mass of both photons. Even though all photons have zero (0) rest mass, they all have mass, m and energy E, while moving at c, (the velocity of light) Photon energy is Planck's constant divided by 2π, (denoted by $\hbar$) multiplied by its angular frequency, ω or $E = \hbar\omega$. Its energy is also its mass multiplied by the speed of light squared, c^2 or $E = mc^2$. In other words the energy, E of a photon may be expressed in two ways as

(2.0A) $E = mc^2 = \hbar\omega$

which relates its mass and angular frequency by equation

(2.0B) $m = \hbar\omega/c^2$

Consider for simplicity that both photons have the same frequency and thus, the same mass, m or

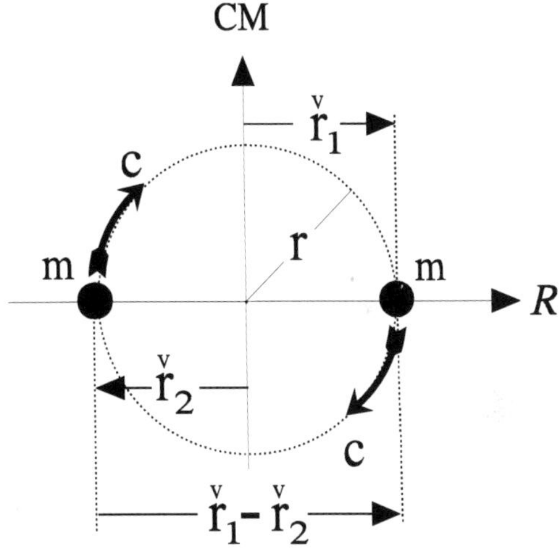

Figure 2 Center of mass (CM) of two identical photons influenced by each other's gravitational field.

(2.01) $m = m_1 = m_2$

Figure 2 shows this system of two photons. Note that since both masses are the same

(2.02) $|\check{r}_1| = r$ and

(2.03) $|\check{r}_2| = r$

where $|\check{r}|$ denotes the magnitude of the vector, $\check{r}$ and

(2.04) $r > 0$

is the radius of the circle. The distance, r_{12} between both photons is

(2.05) $r_{12} = |\check{r}_1 - \check{r}_2| = 2r$

Photon Dynamics

Each photon moves at the same speed of light, c or

(2.06) $v_1 = v_2 = v = c$

The velocity of light, c can always be expressed as

(2.07) $c = \omega\lambda$

where ω is the photon's angular frequency and $\bar{\lambda}$ is related to its wavelength, λ by

$$(2.08)\ \bar{\lambda} = \lambda/(2\pi)$$

In general the relativistic kinetic energy T, is defined as

$$(2.09)\ T = (m - m_0)c^2$$

where m_0 is known as the rest (when m is not moving) mass. Moreover, the total dynamic energy, E of a photon is given by

$$(2.1)\ E = mc^2$$

since a photon has zero rest mass ($m_0 = 0$) and thus, by equations (2.09) and (2.1), its kinetic energy and total energy is

$$(2.2)\ T = E = mc^2$$

Is this coupling ever observed? Ask the author.

Point Particle Origin of Planck Length

Plugging into equation

(1.2) $m_1v_1^2/r_1 = Gm_1m_2/r_{12}^2 = m_2v_2^2/r_2$

from equations

(2.01) $m = m_1 = m_2$

(2.05) $r_{12} = |\check{r}_1 - \check{r}_2| = 2r$ and

(2.06) $v_1 = v_2 = v = c$ we get

(2.3) $mc^2/r = Gm^2/(2r)^2$ which implies

(2.4) $r = Gm/(4c^2)$

From equation

(2.0B) $m = \hbar\omega/c^2$

and equation

(2.07) $c = \omega\lambda\!\!\!-$

one may deduce that

(2.5) $m = \hbar/(c\lambda\!\!\!-)$

which means that equation (2.4) becomes

(2.6) $r = G\hbar/(4c^3\lambda\!\!\!-)$

but the optical path length of the photon is $2\pi r$ which must be an integral number of wavelengths so,

(2.7) $n\lambda\!\!\!- = 2\pi r$ or

(2.8) $r = n\lambda\!\!\!-$

where n = 1, 2, 3, .. so plugging equation (2.8) into equation (2.6) we get

(2.9) $n\lambda\!\!\!\!-^2 = G\hbar/(4c^3)$

which yields

(2.10) $\lambda\!\!\!\!-_n = L_p/(2n^{1/2})$

where the Planck length, L_p is defined to be

(2.11) $L_p = (G\hbar/c^3)^{1/2}$

Using equation

(2.08) $\lambda\!\!\!\!- = \lambda/(2\pi)$

equation (2.10) reduces to

(2.12) $\lambda_n = \pi L_p/n^{1/2}$

Notice that the photon's wavelength has a maximum which occurs at n = 1. Thus, we have

(2.13) $\lambda_1 = \pi L_p$

for the first energy state wavelength, λ_1

Point Particle Origin of Planck Mass

Plugging equation

(2.10) $\lambdabar_n = L_p/(2n^{1/2})$ and

(2.11) $L_p = (G\hbar/c^3)^{1/2}$

into equation

(2.5) $m = \hbar/(c\lambdabar)$

results in

(2.14) $m_n = 2n^{1/2}(c\hbar/G)^{1/2}$

but since the Planck Mass has been defined as

(2.15) $M_p = (c\hbar/G)^{1/2}$

then each photon's mass is,

(2.16) $m_n = 2n^{1/2}M_p$

Note that the mass of each photon is a minimum when n = 1, thus

(2.17) $m_1 = 2M_p$

Point Particle Photon Energy

Recall the kinetic energy, T for each photon is given by equation

(2.2) $T = E = mc^2$

so that by equation (2.16) becomes

(2.18) $E_n = m_nc^2 = 2n^{1/2}M_pc^2$

since its rest mass is zero. Note that each photon's kinetic energy has a minimum when n = 1 given as

this is a huge energy!

(2.18.1) $E_{min} = 2M_pc^2$

Point Particle Origin of Planck Time

The period, τ (time of a round trip) of either photon is

(2.19) $\tau = 2\pi r/c$

and because of equation

(2.8) $r = n\bar{\lambda}$ becomes

(2.20) $\tau_n = 2\pi n\bar{\lambda}/c$

and plugging in for

(2.10) $\bar{\lambda}_n = L_p/(2n^{1/2})$

equation (2.20) reduces to

(2.21) $\tau_n = \pi n^{1/2} L_p/c$

However, the Planck time is defined to be

(2.22) $T_p = L_p/c = (G\hbar/c^5)^{1/2}$

so equation (2.21) becomes

(2.23) $\tau_n = \pi n^{1/2} T_p$

Point Particle
Photon Angular Frequency

The angular frequency ω_n, of each photon is related to its wavelength by

(2.07) $c = \omega\lambdabar$

which makes

(2.07.1) $\omega_n = c/\lambdabar_n$

and by equation

$$(2.10)\ \lambda\!\!\!^{-}_n = L_p/(2n^{1/2})$$

equation (2.07.1) becomes

$$(2.07.2)\ \omega_n = 2n^{1/2}c/L_p$$

and since by

$$(2.22)\ T_p = L_p/c = (G\hbar/c^5)^{1/2}$$

reduces equation (2.07.2) to

$$(2.23.1)\ \omega_n = 2n^{1/2}/T_p$$

Note that the angular frequency of the photon has a minimum at $n = 1$ so we have

$$(2.24)\ \omega_1 = 2/T_p$$

where ω_1 is the first energy state angular frequency.

Point Particle Total Photon Energy

Since the kinetic energy of each photon is

(2.18) $E_n = m_n c^2 = 2n^{1/2} M_p c^2$

Then, both photons have total kinetic energy

(2.25) $T_T = 2E = 4n^{1/2} M_p c^2$

Recall the classic gravitational potential energy of both photons is given by

(1.1.1) $V = - Gm_1 m_2 / r_{12}$

Plugging into equation (1.1.1) from equations

(2.16) $m_n = 2n^{1/2} M_P$

(2.01) $m = m_1 = m_2$

(2.05) $r_{12} = |\check{r}_1 - \check{r}_2| = 2r$

equation (1.1.1) becomes

(2.1.2) $V_n = -\,4GnM_p^2/2r$

Plugging into equation (2.1.2) from equations

(2.8) $r = n\lambda\!\!\!^{-}$ and

(2.10) $\lambda\!\!\!^{-} = L_p/(2n^{1/2})$ yields

(2.1.3) $V_n = -\,4n^{1/2}GM_p^2/L_p$

but since $GM_p/L_p = c^2$ equation (2.1.3) reduces to

(2.1.4) $V_n = -\,4n^{1/2}M_pc^2$

Total kinetic energy by equation (2.25) is $4n^{1/2}M_pc^2$

which makes the total energy, E_T of both photons

(2.26) $E_{Tn} = T_T + V = 4n^{1/2}M_pc^2 - 4n^{1/2}M_pc^2 = 0$

for all values of n including n = 1 making

(2.26.1) $E_{T1} = 0$

Point Particle Orbital Angular Momentum

By definition the vector orbital angular momentum, Ĺ is

(2.27) Ĺ = ŕ x mύ

where ŕ and mύ are the vector position and vector momentum respectively. Applying this to one photon we get

(2.27.1) ℓ = |Ĺ| = mcr

Plugging into equation (2.27) from equations

(2.16) $m_n = 2n^{1/2}M_p$

(2.8) $r = n\lambda\!\!\!^{-}$ and

(2.10) $\lambda\!\!\!^{-}_n = L_p/(2n^{1/2})$

yields

(2.28) $\ell_n = (2n^{1/2}M_p)cn(L_p/(2n^{1/2}))$

$= nM_pcL_p$

$= nc(c\hbar/G)^{1/2}(G\hbar/c^3)^{1/2}$

$= n\hbar$

where again n = 1, 2, 3, . which is reasonable since classically, orbital angular momentum has integral Planck values. Note that the minimum orbital angular momentum, ℓ_1 for each first energy state photon (n = 1) is

(2.28.1) $\ell_1 = \hbar$

Chapter 3
Wave Mechanical Preliminaries

The following equation

$$(3.0)\ -\hbar^2\nabla^2\Psi(r,t)/(m+m_0) + V(r)\Psi(r,t) = E_T\Psi(r,t)$$

was derived in a book entitled "An Alternative Lorentz Invariant Relativistic Wave Equation" by the author.

Meaning of The Alternative Relativistic Wave Equation

This equation is Lorentz invariant and properly describes the wave nature of systems including those having zero rest mass, m_0 as well as relativistic mass m. Prior to this equation, there was no way to describe a relativistic quantum mechanical system in which zero rest mass particles move under the influence of a central field of force. In equation

$$(3.0)\ -\hbar^2\nabla^2\Psi(r,t)/(m+m_0) + V(r)\Psi(r,t) = E_T\Psi(r,t)$$

m_0 is the rest mass and m is its mass in motion assumed to be a function of its velocity according to

$$(3.1)\ m = m_0(1-(v/c)^2)^{-1/2}$$

featured in Einstein's special theory of relativity.

Referring to equation (3.0), the system can be moving under the influence of a central force derivable from a potential V(r). The wave function, $\Psi(r,t)$ is assumed to be a function of position, r and time, t. When describing stationary states with total energy, E_T, the wave function is assumed to be a function of position only. Again, Planck's constant divided by 2π is denoted by $\hbar$. The Del operator is denoted by

$$(3.2)\ \nabla = (\partial/\partial x, \partial/\partial y, \partial/\partial z)$$

and Laplacian

$$(3.3)\ \nabla^2 = \nabla \bullet \nabla = (\partial^2/\partial x^2, \partial^2/\partial y^2, \partial^2/\partial z^2)$$

where • denotes the normal vector dot product. Equation

$$(3.0)\ -\hbar^2\nabla^2\Psi(r,t)/(m+m_0) + V(r)\Psi(r,t) = E_T\Psi(r,t)$$

may be rewritten as

$$(3.4)\ \hat{H}\Psi(r,t) = E_T\Psi(r,t)$$

where $\hat{H}$ is the relativistic Hamiltonian operator given by

(3.5) $\hat{H} = -\hbar^2\nabla^2/(m + m_0) + V(r)$

Hamiltonian Operator For A System of Two Particles

Beginning on page 173 of David Park's book entitled "Introduction to the Quantum Theory" (See reference section) a general two particle wave equation is developed. Following this philosophy, the resulting Hamiltonian operator is

(3.6) $\hat{H}_T = \check{T}_1 + \check{T}_2 + V(\check{r}_1 - \check{r}_2)$

where $V(\check{r}_1 - \check{r}_2)$ is the potential energy between the two particles at vector positions $\check{r}_1$ and $\check{r}_2$. $\check{T}_1$ and $\check{T}_2$ are the kinetic energy operators for both particles and $\hat{H}_T$ is the Hamiltonian operator for the combined system. This implies

(3.7) $\hat{H}_T\Psi_T(r_1,r_2) = E_T\Psi_T(r_1,r_2)$

where E_T is the total energy eigenvalue for the possible stationary energy states of both photons.

The wave function representing both particles is $\Psi_T(r_1,r_2)$ and it is assumed that

(3.8) $\Psi_T(r_1,r_2) = \Psi(r_1)\Psi(r_2)$

Kinetic Energy Operators

Furthermore, by the form of equation

(3.0) $-\hbar^2\nabla^2\Psi(r,t)/(m+m_0) + V(r)\Psi(r,t) = E_T\Psi(r,t)$

it is assumed that the kinetic energy operators obey equations

(3.9) $\check{T}_1 = -\hbar^2\nabla_1^{\,2}/(m_1 + m_{10})$ and

(3.10) $\check{T}_2 = -\hbar^2\nabla_2^{\,2}/(m_2 + m_{20})$

where m_1, m_{10}, m_2 and m_{20} are the masses and rest masses of the two systems respectively. Moreover, the two Laplacians of equations (3.9) and (3.10) are defined by

(3.11) $\nabla_1^2 = (\partial^2/\partial x_1^2, \partial^2/\partial y_1^2, \partial^2/\partial z_1^2)$ and

(3.12) $\nabla_2^2 = (\partial^2/\partial x_2^2, \partial^2/\partial y_2^2, \partial^2/\partial z_2^2)$

where the position (see Figure 1) of particle one is

(3.13) $r_1 = (x_1, y_1, z_1)$ and

simultaneously the position of particle two is

(3.14) $r_2 = (x_2, y_2, z_2) = (-x_1, -y_1, -z_1) = -r_1$

Alternate Form of Laplacian Operator

On pages 516 – 518 of David Park's book entitled "Introduction to the Quantum Theory" (See reference section) it can be deduced that the Laplacian operator is

(3.15) $\nabla^2 = d^2/dr^2 + (2/r)d/dr - \acute{L}^2/r^2$

where

$$(3.15.1)\ \acute{L}^2 = -[(1/(\sin\theta))(\partial/\partial\theta(\sin\theta\partial/\partial\theta) + (1/\sin^2\theta)(\partial^2/\partial\varphi^2)]$$

$\acute{L}^2$ is the square of the orbital angular momentum operator, θ is the angle that the projection of the radial distance, r onto the xy plane, makes with the x axis and φ is the angle the radial distance, r makes with the z axis (normal spherical coordinates). It turns out that the eigenvalues of $\acute{L}^2$ performed on a normalized spherical wave function described by r, θ and φ, Ψ(r,θ,φ) is

$$(3.15.2)\ \acute{L}^2\Psi(r,\theta,\varphi) = \ell(\ell+1)\Psi(r,\theta,\varphi)$$

where the orbital angular momentum quantum number, ℓ can only take on non–negative integer values (0, 1, 2, ..). If the wave function depends only on r, then equation (3.15.2) reduces to

$$(3.15.3)\ \acute{L}^2\Psi(r) = 0$$

This completes the development of the preliminary tools, which will be used to describe a system of two photons trapped in each other's gravitational field using relativistic wave mechanics.

why wouldn't the photons be absorbed by materials, i.e. real materials

There has to be some weakly interacting but massive particles

Chapter 4
Wave Mechanical System

Let us begin by utilizing equations

(3.6) $\hat{H}_T = \check{T}_1 + \check{T}_2 + V(\check{r}_1 - \check{r}_2)$

(3.7) $\hat{H}_T\Psi(r_1,r_2) = E_T\Psi(r_1,r_2)$

(3.9) $\check{T}_1 = -\hbar^2\nabla_1{}^2/(m_1 + m_{10})$

(3.10) $\check{T}_2 = -\hbar^2\nabla_2{}^2/(m_2 + m_{20})$

Two Photon Wave System

Recalling that the rest mass for both photons is zero, equation (3.7) in the center of mass system for two identical photons as in Figure 2 becomes

$$(4.0)\ -\hbar^2\nabla_1^{\,2}\Psi_T/m + -\hbar^2\nabla_2^{\,2}\Psi_T/m + V\Psi_T = E_T\Psi_T$$

where the total wave function, Ψ_T has joint probability amplitude written as $\Psi(r_1,r_2) = \Psi(r_1)\Psi(r_2)$ and potential energy function $V(\check{r}_1 - \check{r}_2)$ written as V. The stationary energy eigenvalues of both photons is E_T. Note that $m = m_1 = m_2$ for identical photons. Equation (4.0) may be further simplified by realizing that since

$(4.1)\ \check{r}_2 = -\check{r}_1$ implies that

$$(4.1.1)\ \nabla_2 = -\nabla_1$$

so that the Laplacians for both particles are equal or

(4.2) $\nabla_1^2 = \nabla_2^2 = \nabla^2$

The mass of each identical photon is

(2.0B) $m = \hbar\omega/c^2$

Plugging equation (4.2) and (2.0B) into (4.0) yields

(4.3) $-2\hbar c^2\nabla^2\Psi_T/\omega + V\Psi_T = E_T\Psi_T$

Potential Energy Term

Recall that the gravitational potential energy is

(1.1.1) $V = -Gm_1m_2/r_{12}$

and because of

(2.05) $r_{12} = |\check{r}_1 - \check{r}_2| = 2r$

and since by

(2.01) $m = m_1 = m_2$

reduces equation (1.1.1) to

(4.4) $V = -\,Gm^2/(2r)$

and plugging in from equation (2.0B) can also be expressed as

(4.5) $V = -G\hbar^2\omega^2/(2rc^4)$

Relativistic Wave Equation of Two Photons

Plugging equation (4.5) into equation

(4.3) $-2\hbar c^2\nabla^2\Psi_T/\omega \; + V\Psi_T = E_T\Psi_T$

yields

(4.6) $-2\hbar c^2\nabla^2\Psi_T/\omega \; + -G\hbar^2\omega^2/(2rc^4)\Psi_T = E_T\Psi_T$

which simplifies into

$$(4.7)\ \nabla^2\Psi_T + G\hbar\omega^3/(4rc^6)\Psi_T + \omega E_T/(2\hbar c^2)\Psi_T = 0$$

If one plugs the Laplacian from equation

$$(3.15)\ \nabla^2 = d^2/dr^2 + (2/r)d/dr - \acute{L}^2/r^2$$

into equation (4.7) yields

$$(4.8)\ d^2\Psi_T/dr^2 + (2/r)d\Psi_T/dr - \acute{L}^2\Psi_T/r^2 + (G\hbar\omega^3/(4rc^6))\Psi_T + (\omega E_T/(2\hbar c^2))\Psi_T = 0$$

It will sometimes be convenient to write this equation in terms of the mass of a photon in motion and using the fact that $m = \hbar\omega/c^2$, equation (4.8) becomes

$$(4.9)\ d^2\Psi_T/dr^2 + (2/r)d\Psi_T/dr - \acute{L}^2\Psi_T/r^2 + (G\hbar m^3/(4r\hbar^2))\Psi_T + (mE_T/(2\hbar^2))\Psi_T = 0$$

Note the wave function, Ψ represent both photons and is given by

$$(4.10)\ \Psi_T = \Psi_1\Psi_2 = \Psi_T(r_1,r_2,\theta_1,\theta_2,\varphi_1,\varphi_2)$$

In the center of mass system of both photons, the model in spherical coordinates allows for the following observations

$$(4.10.1)\ r_1 = r_2 = r \text{ and}$$

$$(4.10.2)\ \theta_1 = \theta_2 - \pi = \theta \text{ and}$$

$$(4.10.3)\ \varphi_1 = \varphi_2 - \pi = \varphi$$

so that the wave function can be simply written as

$$(4.10.4)\ \Psi_T(r_1,r_2,\theta_1,\theta_2,\varphi_1,\varphi_2) = \Psi_T(r,\theta,\varphi) = \Psi_T$$

The two equivalent forms can be written as equation

(4.8) $d^2\Psi_T/dr^2 + (2/r)d\Psi_T/dr \; - \acute{L}^2\Psi_T/r^2 +$

$$(G\hbar\omega^3/(4rc^6))\Psi_T + (\omega E_T/(2\hbar c^2))\Psi_T = 0$$

or as equation

(4.9) $d^2\Psi_T/dr^2 + (2/r)d\Psi_T/dr \; - \acute{L}^2\Psi_T/r^2 +$

$$(G\hbar m^3/(4r\hbar^2))\Psi_T + (mE_T/(2\hbar^2))\Psi_T = 0$$

Thus, either of these equations represent the relativistic wave equation describing two identical photons moving under the influence of each other's gravitational field. Ψ_T represents the total wave function of both photons.

The next chapter will begin dealing with its solution.

Chapter 5
Towards the Solution

Equation

$$(4.9)\ d^2\Psi_T/dr^2 + (2/r)d\Psi_T/dr - \acute{L}^2\Psi_T/r^2 + (G\hbar m^3/(4r\hbar^2))\Psi_T + (mE_T/(2\hbar^2))\Psi_T = 0$$

is the total wave function, Ψ_T expressible in spherical coordinates, and representing two photons trapped in each other's gravitational field.

Factors of Total Wave Function

This equation may be solved by assuming that the wave function, Ψ_T can be written in terms of the product of a radial part, $\Psi(r)$ and an angular part as

(5.0) $\Psi_T(r,\theta,\varphi) = \Psi(r)\mathcal{Y}_\ell^m(\theta,\varphi)$

Spherical Harmonics

where $\mathcal{Y}_\ell^m(\theta,\varphi)$ are the normalized spherical harmonics of degree ℓ defined by

(5.1) $\mathcal{Y}_\ell^m(\theta,\varphi) = \{[(2\ell + 1)((\ell - m)!)]/[(4\pi)((\ell + m)!)]\}^{1/2} P_\ell^m e^{im\varphi}$

Legendre Functions

The $P_\ell^m(u) = P_\ell^m(\cos\theta)$ are known as the associated Legendre functions defined as

(5.2) $P_\ell^m(u) = (1/(2^\ell \ell!))(-1)^m(1 - u^2)^{m/2}(d^{\ell+m}/du^{\ell+m})(u^2 - 1)^\ell$

Note that ℓ and m can only take on integer values with m taking on values of $-\ell$, $-\ell+1$, .., 0, $\ell+1$,.. ℓ and ℓ taking on values of 0, 1, 2, .. etc.

Orbital Angular Momentum Operators

The square of the orbital angular momentum operator , $\acute{L}^2$ satisfies

(5.3) $\acute{L}^2 \mathcal{Y}_\ell^m(\theta,\varphi) = \ell(\ell+1)\mathcal{Y}_\ell^m(\theta,\varphi)$

and the z component of the angular momentum operator, $\acute{L}_Z$ satisfies

(5.4) $\acute{L}_Z \mathcal{Y}_\ell^m(\theta,\varphi) = m\mathcal{Y}_\ell^m(\theta,\varphi)$

Radial and Angular Components of Total Wave Function

When equation (5.0) $\Psi_T(r,\theta,\varphi) = \Psi(r)\mathcal{Y}_\ell^m(\theta,\varphi)$ is substituted into equation

(4.9) $d^2\Psi_T/dr^2 + (2/r)d\Psi_T/dr - \acute{L}^2\Psi_T/r^2 + (G\hbar m^3/(4r\hbar^2))\Psi_T + (mE_T/(2\hbar^2))\Psi_T = 0$

a pair of equations are produced namely equation

$$(5.3)\ \acute{L}^2 \mathcal{Y}_\ell^{m}(\theta,\varphi) = \ell(\ell+1)\mathcal{Y}_\ell^{m}(\theta,\varphi)$$

and a radial equation

$$(5.5)\ d^2\Psi/dr^2 + (2/r)d\Psi/dr - \ell(\ell+1)\Psi/r^2 + (G\hbar m^3/(4r\hbar^2))\Psi + (mE_T/(2\hbar^2))\Psi = 0$$

This last equation contains the energy eigenvalue, E_T or binding energy, $-E_T$ of both trapped photons, thus, its solution should yield both the energy of each photon as well as these eigenvalues.

Normalizable Total Wave Function

The total wave equation

$$(4.9)\ d^2\Psi_T/dr^2 + (2/r)d\Psi_T/dr - \acute{L}^2\Psi_T/r^2 + (G\hbar m^3/(4r\hbar^2))\Psi_T + (mE_T/(2\hbar^2))\Psi_T = 0$$

may be rewritten utilizing

(3.15) $\nabla^2 = d^2/dr^2 + (2/r)d/dr - \acute{L}^2/r^2$

so that equation (4.9) becomes

(4.9.1) $\nabla^2\Psi_T + (G\hbar m^3/(4r\hbar^2))\Psi_T + (mE_T/(2\hbar^2))\Psi_T = 0$

It turns out that if this equation has the same form as in Korn's *Mathematical Handbook for Scientists and Engineers* (see reference section), page 319, equation

(10.4–33) $\nabla^2\Phi + (2K_1/r - K_2{}^2)\Phi = 0$

where K_1 and K_2 are constants and $K_1/K_2 = n$, a natural number, then Korn says it has a normalizable solution given by equation

(10.4–34) $\Phi(r,\theta,\varphi) = r^j e^{-K_2 r} L(2K_2 r)^{2j+1}{}_{n+j} Y_j(\theta,\varphi)$

where the $L(2K_2r)^{2j+1}{}_{n+j}$ are known as the associated Laguerre functions of $(2K_2r)$ defined by

$$(5.6)\ L(x)_a{}^b = (x^{-b}e^x/a!)d^a/dx^a(e^{-x}x^{a+b})$$

where a and b are both integers and $Y_j(\theta,\varphi)$ is what Korn calls a spherical surface harmonic of degree j (see page 317). It turns out that these were previously identified as the postulated solutions of the angular equation

$$(5.3)\ \acute{L}^2\mathcal{Y}_\ell{}^m(\theta,\varphi) = \ell(\ell+1)\mathcal{Y}_\ell{}^m(\theta,\varphi)$$

where the functions of equation

$$(5.1)\ \mathcal{Y}_\ell{}^m(\theta,\varphi) = \{[(2\ell+1)((\ell-m)!)]/[(4\pi)((\ell+m)!)]\}^{1/2}P_\ell{}^m e^{im\varphi}$$

were defined as a spherical harmonic of degree ℓ.

Unnormalized Trial Solutions to the Radial Equation

Using this information as a guide, some particularly simple unnormalized radial wave functions that satisfy equation

$$(5.5)\ d^2\Psi/dr^2 + (2/r)d\Psi/dr - \ell(\ell+1)\Psi/r^2 + (G\hbar m^3/(4r\hbar^2))\Psi + (mE_T/(2\hbar^2))\Psi = 0$$

are postulated to be of the form

$$(5.7)\ \Psi_n = e^{-k_n r}(k_n r)^{n-1} L(k_n r)^{2\ell+1}{}_{n-(\ell+1)}$$

where k_n is a constant dependent on n (known as the energy quantum number). Note that n is a natural number and represents the number of photon wavelengths in one orbit (optical path length) which is the circumference of a circle of radius r as in figure 2. This is expressed by equation

$$(2.8)\ r = n\lambda\!\!\!^{-}$$

and can be expressed in other ways such as

$$(5.8)\ r = n\bar{\lambda} = n\lambda/(2\pi) = nc/\omega = nc\hbar/\hbar\omega = n\hbar/mc$$

where λ is the photon wavelength, ω is its angular frequency and m is its mass in flight. As usual, c is the speed of light in vacuum and $\hbar$ is planck's constant divided by 2π. The orbital angular momentum quantum number, ℓ can take on integer values of 0, 1, 2, … (n–1).

The next chapter is devoted to establishing a procedure for finding energy eigenstates and eigenvalues.

Chapter 6
First Energy Eigenstate

Using equation

$$(5.7)\ \Psi_n = e^{-k_n r}(k_n r)^{n-1} L(k_n r)^{2\ell+1}{}_{n-(\ell+1)}$$

radial wave functions may be specified.

First Energy
Trial Wave Function

Setting

$$(6.0)\ \ n = 1$$

and

(6.1) $\ell = 0$

equation (5.7) becomes

(6.2) $\Psi_1 = e^{-k_1 r}$

where Ψ_1 means the first (n = 1) energy eigenfunction. Substituting Ψ_1 into equation

(5.5) $d^2\Psi/dr^2 + (2/r)d\Psi/dr - \ell(\ell+1)\Psi/r^2 + (Gm^3/(4r\hbar^2))\Psi + (mE_T/(2\hbar^2))\Psi = 0$

results in

(6.3) $k_1^{\,2} - 2k_1/r + Gm_1^{\,3}/(4r\hbar^2) + m_1E_{T1}/(2\hbar^2) = 0$

where the first energy state is E_{T1} and the first energy state mass of either photon is m_1 and k_1 is a constant.

Results of Trial Solution

By inspection, this equation is satisfied if both

(6.3.1) $k_1^{\ 2} + m_1E_{T1}/2\hbar^2 = 0$

and

(6.3.2) $-2k_1/r + Gm_1^{\ 3}/(4r\hbar^2) = 0$

which reduces to

(6.3.2.1) $-2k_1 + Gm_1^{\ 3}/(4\hbar^2) = 0$

Solving equation (6.3.1) for E_{T1} yields

(6.4) $E_{T1} = -k_1^{\ 2}2\hbar^2/m_1$

and solving for k_1 in equation (6.3.2.1) yields

(6.5) $k_1 = Gm_1^{\ 3}/(8\hbar^2)$

plugging this into (6.4) and simplifying, we get:

(6.6) $E_{T1} = -G^2 m_1^{\ 5}/(32\hbar^2)$

but by the definition of the total energy for two identical photons yields:

(6.7) $2m_1c^2 + V_1 = E_{T1}$

and plugging in V_1 from equation

(4.4) $V = -Gm^2/(2r)$

becomes

(6.7.1) $V_1 = -Gm_1^{\ 2}/(2r)$

Plugging equations (6.6) and (6.7.1) into equation (6.7) yields

(6.8) $2m_1c^2 - Gm_1^{\ 2}/(2r) = -G^2 m_1^{\ 5}/(32\hbar^2)$

but by equation

$$(5.8)\ r = n\lambda\!\!\!^{-} = n\lambda/(2\pi) = nc/\omega = nc\hbar/\hbar\omega = n\hbar/mc$$

so plugging $r = n\hbar/m_1c$ into equation (6.8), dividing by m_1c^2 and rearranging terms we obtain equation

$$(6.10)\ G^2m_1^{\,4}/(32\hbar^2c^2) - Gm_1^{\,2}/(2n\hbar c) + 2 = 0$$

Multiplying through by $32\hbar^2c^2$ and dividing by G^2 yields

$$(6.12)\ m_1^{\,4} - m_1^{\,2}16\hbar c/(Gn) + 64\hbar^2c^2/G^2 = 0$$

and by the definition of the Planck mass M_P as

$$(6.13)\ M_P = (\hbar c/G)^{1/2}$$

equation (6.12) reduces to

$$(6.12.1)\ m_1^{\,4} - m_1^{\,2}16M_P^{\,2}/n + 64M_P^{\,4} = 0$$

which is a quadratic whose solution is

(6.14) $m_1^2 = 8M_P^2[1 \pm (1-n^2)^{1/2}]/n$

Clearly, the integer n cannot be zero since that would make the photon mass, m_1 infinite. Moreover, n cannot be greater than 1, since that would make the mass complex. Thus, the only possibility is that n = 1 which is consistent with the initial choice of the wave function

(6.2) $\Psi_1 = e^{-k_1 r}$

First Energy Photon Mass

and equation (6.14) becomes

(6.14.1) $m_1^2 = 8M_P^2$

which means

(6.15) $m_1 = 8^{1/2}M_P$

First Energy Photon Frequency

By equation

(2.0B) $m = \hbar\omega/c^2$

the angular frequency of either photon is

(6.15.01) $\omega_1 = 8^{1/2}/T_P$

where T_P is the Planck time defined as

(6.15.02) $T_P = (\hbar G/c^5)^{1/2}$

Thus, the energy, E_1 of each photon by equation

(2.0B) $m = \hbar\omega/c^2$

is expressed as

(6.15.03) $E_1 = 8^{1/2}M_Pc^2$ or

(6.15.04) $E_1 = \hbar 8^{1/2}/T_P$

First Energy Photon Wavelength

Moreover, the photon wavelength can be calculated using equation

(2.07) $c = \omega ƛ$

Plugging into this from (6.15.01) and solving for $ƛ$ results in

(6.15.05) $ƛ_1 = c/\omega_1 = cT_P/(8)^{1/2} = L_P/(8)^{1/2}$

where L_P is the Planck length defined as

(2.11) $L_p = (G\hbar/c^3)^{1/2} = cT_P$

Since $ƛ_1 = \lambda_1/2\pi$, equation (6.15.05) can be expressed as

(6.15.06) $\lambda_1 = 2\pi L_P/(8)^{1/2}$

Check Using Expectation of Potential Energy

The wave function of equation

$$(6.2)\ \Psi_1 = e^{-k_1 r}$$

may be normalized by assuming that

$$(6.15.2)\ \Psi_{n1} = Ae^{-k_1 r}$$

where A is the normalizing constant satisfying the normalizing condition

$$(6.15.3)\ \int_0^\infty \Psi_{n1}{}^2 d^3r = 1$$

where d^3r is the radial volume element given by

$$(6.15.4)\ d^3r = 4\pi r^2 dr$$

which represents the volume of a spherical shell of radius r, whose infinitesimal width is dr. Equation (6.15.3) then becomes

$$(6.15.5)\quad 4\pi A^2 \int_0^\infty e^{-2k_1 r} r^2 dr = 1$$

Integrating and solving for A yields

$$(6.15.6)\quad A = k_1^{3/2}\pi^{-1/2}$$

which means that the normalized wave function is

$$(6.15.7)\quad \Psi_{n1} = k_1^{3/2}\pi^{-1/2}e^{-k_1 r}$$

Thus the expectation value of 1/r is given by

$$(6.15.8)\quad <1/r> = \int_0^\infty \Psi_{n1}^2(1/r)d^3r$$

$$= 4\pi k_1^3\pi^{-1} \int_0^\infty e^{-2k_1 r} rdr$$

$$= 4k_1^3 \int_0^\infty e^{-2k_1 r} rdr$$

$$= k_1 = Gm_1^3/(8\hbar^2)$$

by equation (6.5) $k_1 = Gm_1^3/(8\hbar^2)$ which means that the potential energy given by equation

(4.4) $V = -Gm^2/(2r)$

becomes

(6.15.9) $V_1 = -\langle 1/r \rangle Gm_1^2/2$

and by equation (6.15.8) becomes

(6.15.10) $V_1 = -\, G^2m_1^5/(16\hbar^2)$

Utilizing equation

(6.7) $2m_1c^2 + V_1 = E_{T1}$

and plugging in the value of the total energy or negative binding energy given by equation

(6.6) $E_{T1} = -G^2m_1^5/(32\hbar^2)$

reduces equation (6.7) to

$$(6.15.11)\ 2m_1c^2 - G^2m_1^{\ 5}/(16\hbar^2) = -G^2m_1^{\ 5}/(32\hbar^2)$$

Dividing by m_1c^2 and multiplying by 32 yields

$$(6.15.12)\ G^2m_1^{\ 4}/(\hbar^2c^2) = 64$$

Taking the square root of both sides yields

$$(6.15.13)\ Gm_1^{\ 2}/(c\hbar) = 8$$

and by the definition of Planck mass

$$(6.13)\ M_P = (\hbar c/G)^{1/2}$$

reduces (6.15.13) to

$$(6.15.14)\ m_1^{\ 2} = 8M_P^{\ 2}$$

Taking the square root of both sides yields

(6.15.15) $m_1 = 8^{1/2}M_P$

which is exactly the same as equation

(6.15) $m_1 = 8^{1/2}M_P$

Note that the result of equation (6.15.15) did not depend on the value of the integer n as did equation (6.15).

First Energy Eigenvalue of Photon Doublet

Recall that the total energy or negative binding energy, E_{T1} of both photons for the n = 1 and ℓ = 0 state was given by equation

(6.6) $E_{T1} = -G^2 m_1^5/(32\hbar^2)$

and because of equation

(6.15) $m_1 = 8^{1/2}M_P$

and equation

(6.15.12) $G^2m_1{}^4/(\hbar^2c^2) = 64$

equation (6.6) becomes:

(6.16) $E_{T1} = -2m_1c^2 = -2\,(8^{1/2})M_Pc^2 = -2\hbar\omega_1$

and since

(6.15.01) $\omega_1 = 8^{1/2}/T_P$

equation (6.16) can be expressed in terms of angular frequency as

(6.17) $E_{T1} = -2\hbar\omega_1 = -(2\hbar)8^{1/2}/T_P$

which is different from the point particle result of the n = 1 point particle energy given by equation

(2.26.1) $E_{T1} = 0$

The next chapter will produce the general solution for both the nth energy eigenstate as well as the nth energy eigenvalue.

Chapter 7
Wave Mechanical Solution

Again using equation

$$(5.7)\ \Psi_n = e^{-k_n r}(k_n r)^{n-1} L(k_n r)^{2\ell+1}{}_{n-(\ell+1)}$$

trial radial wave functions may be specified by choosing the energy quantum number n, from the natural numbers, and then the orbital angular momentum quantum number ℓ knowing that ℓ can range from 0 to n–1. Next, by the procedure of the previous chapter (in which the first energy eigenstate was found for $n = 1$ and $\ell = 0$), any energy eigenstate may be computed.

In fact, the author has carried this through for the next three energy quantum numbers. In each energy case, the orbital angular momentum quantum number was chosen to be one less than the energy quantum number. The resulting eigenstates and energy eigenvalues (negative binding energy) including the first energy eigenstate found in the last chapter) are tabulated as follows;

$$\Psi_1 = e^{-k_1 r} \qquad E_{T1} = -2(8)^{1/2} M_P c^2 \quad (n=1, \ell = 0)$$

$$\Psi_2 = e^{-k_2 r}(k_2 r) \qquad E_{T2} = -8 M_P c^2 \quad (n=2, \ell = 1)$$

$$\Psi_3 = e^{-k_3 r}(k_3 r)^2 \qquad E_{T3} = -4(6)^{1/2} M_P c^2 \quad (n=3, \ell = 2)$$

$$\Psi_4 = e^{-k4r}(k_4 r)^3 \qquad E_{T4} = -8(2)^{1/2} M_P c^2 \quad (n=4, \ell = 3)$$

Note that the binding energy ($-E_{Tn}$) are steadily increasing.

Simplified Energy Eigenstates

It was discovered that once n, the energy quantum number is chosen, the trial wave function of equation

$$(5.7)\ \Psi_n = e^{-k_n r}(k_n r)^{n-1} L(k_n r)^{2\ell+1}{}_{n-(\ell+1)}$$

is greatly simplified by setting the orbital angular momentum quantum number to

$$(5.7.1)\ \ell = n - 1$$

and since by

$$(5.6)\ L(x)_a^{\ b} = (x^{-b}e^{x}/a!)d^{a}/dx^{a}(e^{-x}x^{a+b})$$

makes a = 0, and $L(x)_0^{\ b} = 1$. Thus, the nth energy eigenstate wave function of

$$(5.7)\ \Psi_n = e^{-k_n r}(k_n r)^{n-1}L(k_n r)^{2\ell+1}{}_{n-(\ell+1)}$$

reduces to

$$(7.0)\ \Psi_n = e^{-k_n r}(k_n r)^{n-1} = k_n^{\ n-1}r^{n-1}e^{-k_n r}$$

Substituting this Ψ_n into the radial equation of

$$(5.5)\ \ d^2\Psi/dr^2 + (2/r)d\Psi/dr - \ell(\ell+1)\Psi/r^2 + (Gm^3/(4r\hbar^2))\Psi + (mE_T/(2\hbar^2))\Psi = 0$$

results in

$$(7.1)\ k_n^{n+1}r^{n-1} - 2nr^{n-2}k_n^{\ n} + Gm_n^{\ 3}k_n^{\ n-1}r^{n-2}/(4\hbar^2) + m_nE_{Tn}k_n^{\ n-1}r^{n-1}/(2\hbar^2) = 0$$

where the nth energy state is E_{Tn} and the nth energy state mass of either photon is m_n and k_n is the nth constant. By inspection, this equation is satisfied if both equations

$$(7.2)\ k_n^{\ n+1} + m_nE_{Tn}k_n^{\ n-1}/(2\hbar^2) = 0$$

and

$$(7.3)\ -2nk_n^{\ n} + Gm_n^{\ 3}k_n^{\ n-1}/(4\hbar^2) = 0$$

are both simultaneously satisfied. Equation (7.2) can be reduced to

(7.2.1) $E_{Tn} = -2\hbar^2 k_n^2/m_n$

and equation (7.3) can be reduced to

(7.3.1) $k_n = Gm_n^3/(n8\hbar^2)$

plugging equation (7.3.1) into (7.2.1) and simplifying, yields

(7.4) $E_{Tn} = -G^2 m_n^5/(32n^2\hbar^2)$

Thus, if the nth energy state photon mass, m_n can be found, then the energy eigenstates, E_{Tn} of equation (7.4) can be completely specified.

Wave Origin of Planck Mass

The total energy for two gravitationally interacting photons is

(7.5) $2m_n c^2 + V_n = E_{Tn}$

and using equation (4.4) $V = -Gm^2/(2r)$ produces

(7.6) $V_n = -Gm_n^2/(2r)$

Utilizing equations (7.4) and (7.6), equation (7.5) becomes

(7.7) $2m_nc^2 - Gm_n^2/(2r) = -G^2m_n^5/(32n^2\hbar^2)$

and using

(5.8) $r = n\lambda\!\!\!^{-} = n\lambda/(2\pi) = nc/\omega = nc\hbar/\hbar\omega = n\hbar/mc$

yields

(7.8) $r = n\lambda\!\!\!^{-}_n = n\hbar/m_nc$

which reduces the potential energy term of equation

(7.6) $V_n = -Gm_n^2/(2r)$

to

(7.6.1) $V_n = -Gm_n^3c/(2n\hbar)$

Plugging this into equation (7.7) and simplifying yields

(7.9) $m_n^4 - m_n^2 c\hbar 16n/G + 64n^2c^2\hbar^2/G^2 = 0$

which can be written as

(7.9.1) $(m_n^2 - 8c\hbar n/G)^2 = 0$

which implies

(7.10) $m_n^2 = 8c\hbar n/G = 8nM_P^2$

whose solution is

(7.11) $m_n = (8n)^{1/2}M_P$

Wave Photon Energy

Since the energy of either photon is given by equation

(2.1) $E = mc^2$

and knowing the nth photon mass by equation (7.11) means that the nth photon energy is

(7.11.1) $E_n = (8n)^{1/2}M_Pc^2$

Wave Potential Energy

Plugging in from equation

(7.11) $m_n = (8n)^{1/2}M_P$

reduces the potential energy given by equation

(7.6.1) $V_n = -Gm_n^3c/(2n\hbar)$

to

(7.6.1.1) $V_n = -4(8n)^{1/2}M_Pc^2$

Wave Origin of Planck Time

Utilizing equation

(2.0B) $m = \hbar\omega/c^2$

equation (7.11) corresponds to a photon angular frequency of

(7.12) $\omega_n = (8n)^{1/2}/T_P$

where T_P is the Planck time defined previously in

(6.15.02) $T_P = (\hbar G/c^5)^{1/2}$

Wave Origin of Planck Length

This also corresponds to a photon wavelength which can be computed by equation

(7.8) $r = n\lambda\!\!\!^{-}_n = n\hbar/(m_n c)$ as

(7.13) $\lambda_n = 2\pi\hbar/(m_n c) = 2\pi\hbar/((8n)^{1/2}M_P c)$

which can be simplified by the definition of a Planck length L_P as

(2.11) $L_p = (G\hbar/c^3)^{1/2}$

so that equation (7.13) becomes

(7.13.1) $\lambda_n = \pi L_P/(2n)^{1/2}$

Total Energy Eigenvalues

Plugging m_n from equation (7.11) into equation

(7.4) $E_{Tn} = -G^2 m_n^5/(32n^2\hbar^2)$

yields energy eigenvalues

(7.14) $E_{Tn} = -2(8n)^{1/2} M_P c^2$

This equation reproduces the first four energy eigenvalues (negative binding energy) found by hand and listed at the beginning of this chapter.

The binding energy increases with increasing energy quantum numbers. This is further discussed in the next chapter under the "conclusions" section. Note that all the energy eigenvalues have been found by equation (7.14). However these were uncovered using the simplified unnormalized wave functions (eigenstates of energy) suggested by

$$(7.0)\quad \Psi_n = e^{-k_n r}(k_n r)^{n-1} = k_n^{\,n-1} r^{n-1} e^{-k_n r}$$

Complete Eigenstates

Recall the total wave equation

$$(4.9)\quad d^2\Psi_T/dr^2 + (2/r)d\Psi_T/dr - \acute{L}^2\Psi_T/r^2 + (G\hbar m^3/(4r\hbar^2))\Psi_T + (mE_T/(2\hbar^2))\Psi_T = 0$$

may be rewritten as

$$(7.15)\quad d^2\Psi_T/dr^2 + (2/r)d\Psi_T/dr - \ell(\ell+1)\Psi_T/r^2 + (Gm_n^{\,3}/(4r\hbar^2))\Psi_T + (m_n E_{Tn}/(2\hbar^2))\Psi_T = 0$$

and because of equation

(3.15) $\nabla^2 = d^2/dr^2 + (2/r)d/dr - \acute{L}^2/r^2$

equation (7.15) can be reduced to

(7.16) $\nabla^2\Psi_T + (Gm_n{}^3/(4r\hbar^2))\Psi_T +$

$$(m_nE_{Tn}/(2\hbar^2))\Psi_T = 0$$

Solution Form of Total Wave Equation

Plugging into equation (7.16) from the solution equations

(7.11) $m_n = (8n)^{1/2}M_P$ and

(7.14) $E_{Tn} = -2(8n)^{1/2}M_Pc^2$

yields

(7.16) $\nabla^2\Psi_T +$

$$2n(8n)^{1/2}/(\hbar r)M_Pc\Psi_T - 8nM_P{}^2c^2/(\hbar^2))\Psi_T = 0$$

which can be further simplified in terms of the Planck length L_P as

$$(7.17)\ \nabla^2\Psi_T + [2n(8n)^{1/2}/(L_P r) - 8n/(L_P^2)]\Psi_T = 0$$

Recall the radial function of equation

$$(5.5)\ d^2\Psi/dr^2 + (2/r)d\Psi/dr - \ell(\ell+1)\Psi/r^2 + (Gm^3/(4r\hbar^2))\Psi + (mE_T/(2\hbar^2))\Psi = 0$$

can also be written in terms of these eigenvalues as

$$(7.18)\ d^2\Psi/dr^2 + (2/r)d\Psi/dr - \ell(\ell+1)\Psi/r^2 + [2n(8n)^{1/2}/(L_P r) - 8n/(L_P^2)]\Psi = 0$$

Normalizable Criteria For Total Wave Equation

Again Comparing equation (7.17) with Korn's equation on page 319 of Korn's *Mathematical Handbook for Scientists and Engineers* (see reference section) of

(10.4–33) $\nabla^2\Phi + (2K_1/r - K_2^2)\Phi = 0$

It is observed that

(7.17.1) $K_1 = n(8n)^{1/2}/L_P$ and

(7.17.2) $K_2 = (8n)^{1/2}/L_P$

Note that

(7.17.3) $K_1/K_2 = n$

which is the energy quantum number, and can take on only integral values and so satisfies Korn's requirement that equation

(7.17) $\nabla^2\Psi_T + [2n(8n)^{1/2}/(L_P r) - 8n/(L_P^2)]\Psi_T = 0$

has normalizable eigenstates. Recall that these were given by Korn's equation on page 317 as

(10.4–34) $\Phi(r,\theta,\varphi) = r^j e^{-K_2 r} L(2K_2 r)^{2j+1}{}_{n+j} Y_j(\theta,\varphi)$

which when translated into the terms of equation

$$(7.17)\ \nabla^2\Psi_T + [2n(8n)^{1/2}/(L_P r) - 8n/(L_P^2)]\Psi_T = 0$$

become like equation

$$(5.0)\ \Psi_T(r,\theta,\varphi) = \Psi(r)\mathcal{Y}_\ell^{m}(\theta,\varphi)$$

where the radial function $\Psi(r)$, because of equation (10.4–34) is of the form

$$(7.19)\ \Psi(r) = r^j e^{-K_2 r} L(2K_2 r)^{2j+1}$$

and to repeat for completeness from chapter 5, the $L(2K_2r)^{2\ell+1}{}_{n+\ell}$ are known as the associated Laguerre functions of $(2K_2r)$ defined by

$$(5.6)\ L(x)_a{}^b = (x^{-b}e^{x}/a!)d^a/dx^a(e^{-x}x^{a+b})$$

where a and b are both integers and $\mathcal{Y}_{\ell}^{m}(\theta,\varphi)$ is known as a spherical harmonic (see Korn's page 872, equation (21.8–66)) equivalently defined by equation

(5.1) $\mathcal{Y}_{\ell}^{m}(\theta,\varphi) = \{[(2\ell + 1)((\ell - |m|)!)]/[(4\pi)((\ell + |m|)!)]\}^{1/2} P_{\ell}^{|m|} e^{im\varphi}$

where $P_{\ell}^{m}(u) = P_{\ell}^{m}(\cos\theta)$ are known as the associated Legendre functions defined by

(5.2) $P_{\ell}^{m}(u) = (1/(2^{\ell}\ell!))(-1)^{m}(1 - u^{2})^{m/2}(d^{\ell+m}/du^{\ell+m})(u^{2} - 1)^{\ell}$

Recall that ℓ and m can only take on integer values with m taking on values of $-\ell$, $-\ell+1$, $-\ell+2$.., 0, 1, .. $+\ell$ and ℓ taking on values of 0, 1, 2, .. (n −1).

Reduction of Radial Wave Equation Into Known Solution

Recall the radial equation

$$(7.18)\ \nabla^2\Psi + [2n(8n)^{1/2}/(L_P r) - 8n/(L_P^2)]\Psi = 0$$

Define a constant K_3 as

$$(7.18.1)\ K_3 = 2(8n)^{1/2}/L_P$$

Multiplying equation (7.18) by r and plugging in K_3 yields

$$(7.20)\ r\nabla^2\Psi + [nK_3 - rK_3^2/4]\Psi = 0$$

Substituting in for ∇^2 of equation

$$(3.15)\ \nabla^2 = d^2/dr^2 + (2/r)d/dr - \acute{L}^2/r^2$$

reduces (7.20) to

(7.21) $r(d^2/dr^2)\Psi + 2(d/dr)\Psi +$

$$[nK_3 - rK_3^2/4 - \ell(\ell+1)/r]\Psi = 0$$

Dividing through by K_3 yields

(7.22) $(r/K_3)(d^2/dr^2)\Psi + (2/K_3)(d/dr)\Psi +$

$$[n - K_3r/4 - \ell(\ell+1)/(K_3r)]\Psi = 0$$

and with the substitution

(7.22.1) $z = K_3r$ becomes

(7.23) $z(d^2/dz^2)\Psi + (2)(d/dz)\Psi +$

$$[n - z/4 - \ell(\ell+1)/z]\Psi = 0$$

which is exactly the same form as in Korn's *Mathematical Handbook for Scientists and Engineers* equation

(21.7–10) $z(d^2/dz^2)\Psi + (2)(d/dz)\Psi +$

$$[n - z/4 - j(j+1)/z]\Psi = 0$$

on page 836 (see reference section).

Normalized Radial Eigenfunctions

Korn's radial solutions are given (directly above equation (21.7–10)) by

(b) $\Psi_{nj}(z) = z^{j} e^{-z/2} L(z)^{2j+1}_{n+j}$

with n = 1, 2, 3, … and j = 0, 1, 2, …(n–1) where the associated Laguerre functions, L(z) are again defined (with x = z) by equation

(5.6) $L(x)_a^{\,b} = (x^{-b} e^{x}/a!) d^{a}/dx^{a} (e^{-x} x^{a+b})$

Moreover, Korn now includes the normalizing equation

(21.7–11) $\int_0^{\infty} \Psi_{nj}^{2} x^{2} dx =$

$$2n[(n + j)!]^2/(n - j - 1)!$$

All of the above considerations allow for the general normalization (via 21.7–11) of radial eigenfunctions which are solutions to the radial equation

(7.18) $\nabla^2\Psi + [2n(8n)^{1/2}/(L_P r) - 8n/(L_P{}^2)]\Psi = 0$

Its unnormalized solution can be expressed by utilizing equation

(b) $\Psi_{nj}(z) = z^j e^{-z/2} L(z)^{2j+1}{}_{n+j}$

with

(7.22.1) $z = K_3 r$

as well as the correspondence of $j \rightarrow \ell$ yielding

(7.24) $\Psi^U{}_{n\ell}(K_3 r) = (K_3 r)^\ell e^{-K_3 r/2} L(K_3 r)^{2\ell+1}{}_{n+\ell}$

where the superscript U means unnormalized.

Normalization Constant

Thus, define the normalized eigenfunctions as

(7.25) $\Psi^{N}{}_{n\ell}(K_3r) = A\Psi^{U}{}_{n\ell}(K_3r)$

where A is the normalizing constant and N means normalized. Thus, the criteria for normalization is

(7.26) $\int_0^{\infty} 4\pi(A\Psi^{U}{}_{n\ell})^2(K_3r)^2K_3dr = 1$

which can be rewritten as

(7.27) $4\pi A^2\int_0^{\infty} (\Psi^{U}{}_{n\ell})^2(K_3r)^2K_3dr = 1$

and by equation

(21.7–11) $\int_0^{\infty} \Psi_{nj}{}^2x^2dx =$

$2n[(n+j)!]^2/(n - j - 1)!$

reduces to

(7.28) $4\pi A^2 2n[(n+\ell)!]^2/(n-j-1)! = 1$

so that the normalization constant, A is

(7.28.1) $A = \{(n-j-1)!/(8\pi n[(n+\ell)!]^2)\}^{1/2}$

and the normalized radial eigenfunctions, Ψ^N are

(7.29) $\Psi^N{}_{n\ell}(K_3r) =$

$\{(n-j-1)!/(8\pi n[(n+\ell)!]^2)\}^{1/2}$ X

$\{(K_3r)^\ell e^{-K_3r/2} L(K_3r)^{2\ell+1}{}_{n+\ell}\}$

where X means multiply as in P X Q = PQ.

Normalized Total Eigenfunctions

Moreover, the total (radial and angular) eigenfunctions given by equation

(5.0) $\Psi_T(r,\theta,\varphi) = \Psi(r)\mathcal{Y}_\ell{}^m(\theta,\varphi)$

when normalized reduce to

(7.30) $\Psi_T(r,\theta,\varphi) = \Psi^N(K_3r)\mathcal{Y}_\ell^{\ m}(\theta,\varphi)$

and which when explicitly written out are

(7.31) $\Psi_T(r,\theta,\varphi) = \{(n-j-1)!/(8\pi n[(n+\ell)!]^2)\}^{1/2}$ X

$(K_3r)^\ell e^{-K_3r/2}$ X $(K_3r)^{-(2\ell+1)}e^{K_3r}/(2\ell+1)!$ X

$d^{2\ell+1}/dx^{2\ell+1}(e^{-K_3r}(K_3r)^{3\ell+n+1})$ X

$\{[(2\ell+1)((\ell-|m|)!)]/[(4\pi)((\ell+|m|)!)]\}^{1/2}$ X

$(1/(2^\ell \ell!))(-1)^{|m|}(1-\cos^2\theta)^{|m|/2}$ X

$(d^{\ell+m}/d\cos\theta^{\ell+m})(\cos^2\theta-1)^\ell$

where $K_3 = 2(8n)^{1/2}/L_P$

This completes the solution for the total (radial plus angular) normalized eigenfunctions of a relativistic quantum mechanical system of two photons trapped in each other's gravitational field.

Chapter 8
Summary and Conclusions

Origin of Planck Units

In summary, the main results of this analysis has identified the origin of the Planck length, Planck mass and Planck time by analyzing the simple case of two identical photons interacting via each other's gravitational field. The analysis was done using relativistic point particle mechanics as well as relativistic wave mechanics.

In both approaches, the Planck length, Planck mass, and Planck time emerge as the natural consequence of a gravitational interaction between two photons. The wave approach has yielded non–zero binding energy for these photon doublets.

Physical Quantity	**Point Mechanics**	**Wave Mechanics**
Photon wavelength, λ_n	$\pi L_p/(2n^{1/2})$	$\pi L_P/(8n)^{1/2}$
Photon Mass, m_n	$2n^{1/2}M_p$	$(8n)^{1/2}M_P$
Photon Frequency, ω_n	$2n^{1/2}/T_P$	$(8n)^{1/2}/T_P$
Photon orbital angular momentum, ℓ_n	$n\hbar$	$[\ell(\ell+1)]^{1/2}\hbar$
Photon energy, E_n	$2n^{1/2}M_Pc^2$	$(8n)^{1/2}M_Pc^2$
Potential energy, V_n	$-4n^{1/2}M_Pc^2$	$-4(8n)^{1/2}M_Pc^2$
Total energy, E_{Tn}	0	$-2(8n)^{1/2}M_Pc^2$

Table 8–1 Photon Doublet Physical Results

Table 8–1 is a comparison of the results of point mechanical physical quantities as well as wave mechanical physical quantities for a system of two photons moving in each other's gravitational field.

Quantum numbers n and ℓ can only take on values of $n = 1, 2, ..; \ell = 0, 1, ...(n-1)$.

Total Photon Doublet Angular Momentum

This study should apply to any pair of identical particles having no (zero) rest mass. However, particles exist in two flavors. Those which have an intrinsic spin of half integral multiples of Planck's constant divided by 2π, $\frac{1}{2}\hbar$ are called fermions. They obey Fermi-Dirac statistics. Those which have an intrinsic spin of integral multiples of Planck's constant divided by 2π, $\hbar$ are called bosons. They obey Bose-Einstein statistics.

It is known that the spin of a photon is $1\hbar$ (a boson). If the spin of both photons in the doublet are parallel, then the doublet photon system will have spin angular momentum = $2\hbar$ otherwise, it will have 0 (zero) spin. Recall that the photon's first energy state orbital angular momentum is zero. Thus, the total first energy state angular momentum (spin plus orbital) of these doublet photon systems can either be zero (anti-parallel spins) or $2\hbar$ (parallel spins).

Total Neutrino Doublet Angular Momentum

It is known that the spin of a neutrino is $\frac{1}{2}\hbar$ (a fermion). If the neutrino has zero rest mass, then it could also form a doublet neutrino system. Thus, the spin angular momentum of a first energy state neutrino doublet system must be 0 since the neutrino's spins must be antiparallel because of the Pauli exclusion principle.

Doublet System Statistics

Note that the first energy state of either a boson doublet system or a fermion doublet system as a whole will obey Bose-Einstein statistics. By the Pauli Exclusion Principle, only 2 fermions could occupy a gravitational first energy state while more than 2 bosons could presumably occupy a corresponding gravitational first energy state.

Figure 3 shows the Feynman diagram for these photon doublet systems.

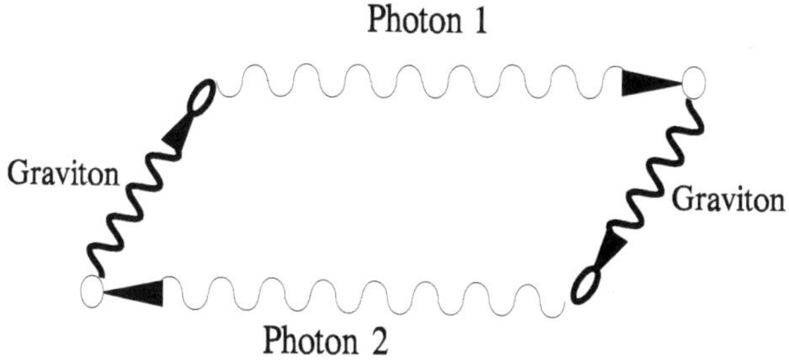

Figure 3 Feynman diagram for two identical photons exchanging gravitions

Candidates For Dark Matter

The photon doublet system described in this manuscript can be generalized to include more than just two photons. Other zero rest mass particles could include fermions as well as bosons. Moreover, the particles could have different masses.

It should also be noted that such boson or fermion systems could have conceivably been born in the early big bang and thus, offer several different forms of candidates for dark matter.

Possible WIMPS

The photon doublet system described herein contains binding energy (negative total energy eigenvalues) that increase with increasing energy quantum numbers. Thus, these photon doublets offer good candidates for weakly interacting massive particles (WIMPS) for relatively large energy quantum numbers. Since this doublet contains no charge, it would very weakly interact.

Possible MACHOS

Moreover, each photon's mass and frequency increase with increasing energy quantum numbers without bound. Thus, these kinds of photon systems also offer good candidates for massive compact halo objects (MACHOS) for relatively large energy quantum numbers.

GLOSSARY

Binding Energy: The amount of energy which must be added to particles bound to each other such that the particles become free (no longer bound).

Bose-Einstein Statistics: Mechanics describing the statistical behavior of a group of Bosons.

Boson: A particle which has its spin angular momentum equal to an integral number of Planck's constant divided by 2π, $\hbar$.

Dark Matter: An invisible form of positive material energy which makes galaxies have more mass than their observable mass. This would explain why galactic stars move at their higher observable speeds than speeds predicted by the observed mass of their associated galaxy.

Del Operator: An operator, ∇ equal to the vector whose xyz components are $\nabla = (\partial/\partial x, \partial/\partial y, \partial/\partial z)$.

Eigenstate: A wave function Ψ (the eigenstate), such that an operator, $\hat{O}$ operating on the wave function, Ψ yields a value, E (the eigenvalue) multiplied by the wave function. If $\hat{O}\Psi = E\Psi$, then Ψ is the eigenstate.

Eigenvalue: If an operator, $\hat{O}$ operating on the wave function, Ψ yields a value, E (the eigenvalue) multiplied by the wave function, then E is the eigenvalue. If $\hat{O}\Psi = E\Psi$, then E is the eigenvalue.

Fermi-Dirac Statistics: Mechanics describing the statistical behavior of a group of Fermions.

Fermion: A particle which has its spin angular momentum equal to a half integral number of Planck's constant divided by 2π, $\hbar/2$.

Feynman Diagram: A standard tool used for the depiction of a quantum mechanical process. Wavy arrows stood for zero rest mass bosons including photons. Solid arrows stand for non-zero rest mass Fermions. Arrows are in the same direction as the flow of time.

Hamiltonian Operator: An operator, $\hat{H}$ which when multiplied by the wave function, Ψ produces the total energy eigenvalue, E. $\hat{H}\Psi = E\Psi$

Helicity: The component of a particles spin angular momentum in the direction of the particle's velocity vector.

Kinetic Energy Operator: An operator, $\check{T}$ which when multiplied by the wave function, Ψ produces the kinetic energy eigenvalue, E_K. $\check{T}\Psi = E_K\Psi$

Laplacian Operator: An operator which is the dot product of the Del = ∇ = $(\partial/\partial x, \partial/\partial y, \partial/\partial z)$ with itself or $\nabla \bullet \nabla = \nabla^2 = (\partial^2/\partial x^2, \partial^2/\partial y^2, \partial^2/\partial z^2)$.

MACHOS: Massive compact halo objects. These are believed to make up much of the energy surrounding galaxies and adding to their mass.

Neutrino: A Fermion particle associated with Beta (electron) radioactivity. See also standard model.

Newton's Universal Gravitational Constant: The force between any two masses is proportional to the product of their masses and inversely proportional to the square of the distance between their centers. The constant of proportionality is Newton's Universal Gravitational Constant, $G = 6.67 \text{ X } 10^{-11}$ newton-meters^2/kilograms^2

Pauli Exclusion Principle: No two Fermions can have all their quantum numbers the same.

Photon: The Boson particle associated with electromagnetic radiation and forces between charges.

Planck Boson Spin: $\hbar$ = 1.0552 X 10^{-34} joule–seconds

Planck's Constant: h = 6.63 X 10^{-34} joule-seconds

Planck's Constant divided by 2π: $h/2\pi = \hbar$ = 1.0552 X 10^{-34} joule-seconds

Planck Fermion Spin: $\hbar/2$ = 5.276 X 10^{-35} joule–seconds

Planck Length: $L_P = (\hbar G/c^3)^{1/2}$ = 1.616 X 10^{-35} meters

Planck Mass: $M_P = (\hbar c/G)^{1/2}$ = 2.177 X 10^{-8} kilograms

Planck Time: $T_P = (\hbar G/c^5)^{1/2}$ = 5.391 X 10^{-44} seconds

Speed of Light in Vacuum: c = 3.00 X 10^{8} meters/second

STANDARD MODEL: All matter is composed of three families of four fermion particles each. The first two members are quarks and the last two members are leptons. These matter families (M) are:

M1 (up quark, down quark, electron, electron anti–neutrino)

M2 (charmed quark, strange quark, muon, muon anti–neutrino)

M3 (top quark, bottom quark, tauon, tauon anti–neutrino)

All field energy is composed of three families of three boson particles each.

The first family F1 includes the gluons responsible for the force that holds the quarks in both neutrons (2 down quarks & 1 up quark) and protons (2 up quarks & 1 down quark) together. The photon is responsible for forces between charges and the graviton is responsible for forces between masses.

The second family F2 includes the 3 pions responsible for holding neutrons and protons of atomic nuclei together.

The third family F3 includes the 3 weakons responsible for forces causing radioactive decay. These three field families (F) are:

F1 (gluon, photon, graviton),

F2 (pi plus, pi minus, pi zero),

F3 (omega plus, omega minus, zeta zero)

All matter particles also have a partner which is called its anti–particle. Anti–particles have the same mass as the particles but the sign of their charge and helicity (if any) is reversed (See basic elementary particles section).

Wave Function: A mathematical function of space and time, which describes the wave behavior of an energy system.

WIMPS: Weakly interacting massive particles. These could include exotic particles or systems of particles.

Fundamental Physical Laws

Preliminary Definitions

I. **Bold** mathematical single letters refer to vectors.

II. The symbol, $i = (-1)^{1/2}$ always occurs as the fourth (time component) of all Einsteinian four dimensional vectors.

III. The symbol, c is the speed of light in vacuum.

1. **Position of an energy system:** Referring to Figure 2, a normal Cartesian coordinate system shows the x,y,z position of the system S at time t.

Newtonian: $\mathbf{r}_N = (x,y,z)$

Einsteinian: $\mathbf{r}_E = (x,y,z,ict)$

2. **Velocity of an energy system:** At time t_2, the position of the system was at position 2 (r_2). Initially at time t_1, the position of the system was at position 1 (r_1). The average velocity of the system is the distance traversed by the system in moving from position 1 to position 2 ($r_2 - r_1$) divided by the time it took for the system to move between the two positions ($t_2 - t_1$). The direction of the velocity is from position 1 to position 2. The instantaneous velocity **v** is realized by letting t_2 approach t_1. Mathematically, the instantaneous velocity of a system is a vector quantity.

$\mathbf{v}$ = lim as $t_2 \rightarrow t_1$ of $[(\mathbf{r}_2 - \mathbf{r}_1)/(t_2 - t_1)]$ or

$\mathbf{v} = d\mathbf{r}/dt$

Newtonian: $\mathbf{v}_N = (v_x, v_y, v_z)$

Einsteinian: $\mathbf{v}_E = (v_x, v_y, v_z, ic)$

3. **Acceleration of an energy system:** At time t_2, the velocity of the system was (v_2). Initially at time

t_1, the velocity of the system was (v_1). The average acceleration of the system is the change in the velocity of the system in going from v_1 to v_2 ($v_2 - v_1$) divided by the time it took for the system to go from v_1 to v_2 ($t_2 - t_1$). The direction of the acceleration is from v_1 to v_2. The instantaneous acceleration **a** is realized by letting t_2 approach t_1. Mathematically, the acceleration of a system is a vector quantity.

$\mathbf{a}$ = lim as $t_2 \rightarrow t_1$ of $[(\mathbf{v_2} - \mathbf{v_1})/(t_2 - t_1)]$ or

$\mathbf{a} = d\mathbf{v}/dt$

Newtonian: $\mathbf{a}_N = (a_x, a_y, a_z)$

Einsteinian: $\mathbf{a}_E = (a_x, a_y, a_z, 0)$

4. **Momentum of an energy system:** The product of the system's mass and its velocity **v** is called its momentum and denoted by **p**. It is a vector quantity having direction **v**.

$\mathbf{p} = m\mathbf{v}$

Newtonian: $\mathbf{p}_N = (p_x, p_y, p_z)$

Einsteinian: $\mathbf{p}_E = (p_x, p_y, p_z, iE/c)$

where E is the total energy = mc^2.

5. **Force on an energy system:** The instantaneous rate of change of a system's momentum with respect to time. Its definition is similar to the definition of velocity. At time t_2 it has momentum 2. At time t_1, it had momentum 1. The average force is the difference in momentum (momentum 2 – momentum 1) divided by the time difference (t_2 – t_1). The instantaneous rate is realized when t_2 approaches t_1.

$\mathbf{F}$ = lim as $t_2 \rightarrow t_1$ of $[((m\mathbf{v})_2 - (m\mathbf{v})_1)/(t_2 - t_1)]$ or

$\mathbf{F} = d(m\mathbf{v})/dt$

Newtonian: $\mathbf{F}_N = m_0 d\mathbf{v}/dt = m_0\mathbf{a}$

where m_0 is the rest mass and **a** is its acceleration.

Einsteinian: $\mathbf{F}_E = d(m\mathbf{v})/dt = m d\mathbf{v}/dt + \mathbf{v} dm/dt$ or

$$\mathbf{F}_E = m_0\mathbf{a}(1-(v/c)^2)^{-3/2} = (c^2/v)dm/dt$$

where **v** is m's velocity and **a** is m's acceleration.

6. **Mass density:** Mass m, per unit volume V.

Average mass density = $\rho_{mavg} = m/V$

Instantaneous mass density = $\rho_m = dm/dV$

7. **Pressure on a surface**: The applied force F, per unit surface area, A.

Average pressure = $P_{avg} = F/A$

Instantaneous pressure = $P = dF/dA$

8. **Angular momentum of an energy system:** Let the vector from the origin to the position of the

system be called the position vector (**r**). The angular momentum of the system (**L**) is then the ordinary vector cross product (**x**), of the position vector with the system's momentum vector (m**v**).

L = **r x** m**v**

9. **Charge density:** Amount of charge q, per unit volume V.

Average charge density = ρ_{qavg} = q/V

Instantaneous charge density = ρ_q = dq/dV

10. **Electrical current:** The instantaneous change of charge q with respect to time.

i = lim as $t_2 \rightarrow t_1$ of $[(q_2 - q_1)/(t_2 - t_1)]$ or

i = dq/dt

11. **Electrical current density:** The electrical current i per unit cross sectional area A of

conductor. The unit vector **u** has a direction of the current i along the conductor perpendicular to the cross sectional area.

$\mathbf{J}_i = \mathbf{u}i/A$ where $i = dq/dt$

In a conductor with conductivity σ_c, the current is in the direction of the electric field **E** and the current density is the product of the conductivity and electric field.

$\mathbf{J}_i = \sigma_c \mathbf{E}$

12. **Mole**: The mass of Avogadro's number of identical molecules or Avogadro's number of identical atoms expressed in grams. One mole of molecules is the molecular weight of the molecule expressed in grams. One mole of atoms is the atomic weight of the atom expressed in grams.

Mechanical Laws

1. **Newton's Laws of Motion:**

1.1 A body will remain at rest or in motion at a constant velocity unless acted on by an unbalanced external force.

1.2 The force on a body is proportional to its acceleration and the constant of proportionality is the rest mass (when the body is at rest), m_0 of the body.

$\mathbf{F} = m_0\mathbf{a}$

Newton was unaware that mass is a function of its velocity.

1.3 The force of one body on a second body is equal and opposite to the force of the second body on the first body or for every action, there is an equal and opposite reaction.

$\mathbf{F}_{12} = -\mathbf{F}_{21}$

1.4 Newton's Universal Law of Gravitation says that any two energy systems having mass attract each other with a force (**F**) proportional to the product of their masses m_1 and m_2 and inversely proportional to the square of the distance (r) between their mass centers. The force is in a direction between the centers of m_1 and m_2, causing them to attract one another and is denoted by the unit vector $\mathbf{r}_u$. G is the constant of proportionality known as Newton's Gravitational Constant. This force is

$$\mathbf{F} = \mathbf{r}_u G m_1 m_2 / r^2$$

deriving the gravitational potential energy, V between m_1 and m_2 as

$$V = -\mathbf{G} m_1 m_2 / r$$

Newtonian: Mass is the cause of the gravitational field.

Einsteinian: Mass energy and momentum warp four dimensional spacetime into a gravitational field.

2. **Quantum Mechanical Laws**:

2.1 An energy system may be described by a wave function. The total energy operator $\hat{H}$ (known as the Hamiltonian) operating on the wave function (Ψ) yields the total energy eigenvalue (E) of the system represented by the wave function. Energy eigenvalues (E) are the allowable energy states that the system may assume. Similarly, other operators operating on the wave function yield other information (such as the spin, momentum, angular momentum, etc.) about the system.

$$\hat{H}\Psi = E\Psi$$

2.2 The square of the wave function $\Psi^*\Psi$, multiplied by a infinitesimal volume d^3r is equal to the infinitesimal probability dP, that a system specified by Ψ, is located within that volume.

$dP = \Psi^* \Psi d^3 r$

2.3 The probability that an energy system represented by the wave function Ψ, is somewhere in all space is unity, which is the basis for a normalized wave function.

$P = \int dP = \int \Psi^* \Psi d^3 r = 1$

3. **The Heisenberg Uncertainty Principle:**

3.1 In an ideal experiment, the product of the standard deviation in the measurement of a system's momentum, Δp and the standard deviation in the measurement of its position, Δr must be greater than a non-zero constant. This constant is Planck's constant divided by 2π, $\hbar$ divided by 2 or $\hbar/2$.

$\Delta p \Delta r \geq \hbar/2 = h/4\pi$

since $\hbar = h/2\pi$ and h is Planck's constant.

This means that an energy system's position and momentum cannot be known simultaneously.

3.2 Another expression of the Heisenberg uncertainty principle is:

$\Delta E \Delta t \geq \hbar/2$

where ΔE is the standard deviation in the measurement of a system's energy and Δt is the standard deviation of the measured times that it had that energy. This means that a system's energy and when it had that energy cannot be known simultaneously.

4. **The energy of a photon (E):** associated with an electromagnetic wave having an angular frequency ω, is the product of Planck's constant divided by 2π, $\hbar$ and its angular frequency, ω. The mass, m_γ is the mass of the photon in flight and c is the speed of light.

$E = \hbar\omega = m_\gamma c^2$

5. **De Broglie's relationship:** which expresses that the wavelength of a particle λ is inversely proportional to its momentum. The constant of proportionality is Planck's constant, *h*.

$\lambda = h/mv$

and is sometimes written as

$\lambdabar = \hbar/mv$

where $\lambdabar = \lambda/(2\pi)$ and $\hbar = h/(2\pi)$

6. **Einstein's Laws of Special Relativity:** The first four relativistic laws are derived by assuming that the velocity of light c, is independent of the velocity of the source of light as well as the velocity of the observer.

6.1 A system's mass m increases if it is moving with a velocity v compared to the velocity of light c, in vacuum. Initially when the system had a velocity of zero, its rest mass is m_0.

$$m = m_0 (1 - (v/c)^2)^{-1/2}$$

6.2 A system's length ℓ decreases if it is moving with a velocity v compared to the velocity of light c, in vacuum. Initially when the system had a velocity of zero, its rest length is ℓ_0. ℓ is in the direction of the velocity.

$$\ell = \ell_0(1 - (v/c)^2)^{1/2}$$

6.3 A system's clock time, t slows (stretches) if it is moving with a velocity v compared to the velocity of light c, in vacuum. Initially when the system was at rest (had a velocity of zero), it had a clock time of t_0.

$$t = t_0 (1 - (v/c)^2)^{-1/2}$$

6.4 The total mechanical energy E of a system containing mass is the product of its mass m and the square of the velocity of light c.

$E = mc^2$

where $m = m_0(1 - (v/c)^2)^{-1/2}$ is dependent on its velocity v. m_0 (rest mass) is its mass when $v = 0$.

6.5 The relativistic kinetic energy T, of a system in motion is the difference (between its mass in motion less its rest mass) times the velocity of light c, squared.

Einsteinian: $T_E = (m - m_0)c^2$

where $m = m_0(1 - (v/c)^2)^{-1/2}$ is dependent on its velocity, v. For small velocities compared to the velocity of light, the Einsteinian kinetic energy reduces to the Newtonian kinetic energy as a first order approximation. For $v \ll c$, $(m - m_0)c^2 \cong (\frac{1}{2})m_0v^2$

Newtonian: $T_N = (½)m_0v^2$

7. **Laws of Thermodynamics**

7.1 The first law of thermodynamics says that within a closed (isolated) system an amount of heat added to the system dQ results in an increase in its internal energy dU and an amount of work done, dW. Usually, dU results in an increase in internal temperature while dW results in a change in volume dV against a constant pressure p. This also means that energy is conserved for a closed system.

$$dQ = dU + dW \quad \text{where} \quad dW = pdV$$

7.2 The second law of thermodynamics says that a change in the entropy dS of a system undergoing a reversible process is defined to be the amount of heat added dQ divided by its temperature T. If the process is irreversible, then the entropy is always greater than the amount of heat added divided by its temperature.

$$dS \geq dQ/T$$

where the equality implies reversibility and the greater than symbol (>) implies irreversibility.

7.3 The perfect gas law says that the gas pressure p multiplied by the volume of gas V is proportional to the number of moles n of gas multiplied by the absolute temperature T of the gas. The constant of proportionality R is known as the universal gas constant.

$$pV = nRT$$

7.4 The fundamental law of heat conduction says that the rate of heat flow dQ/dt across a infinitely thin slab dx of material perpendicular to the surface of the slab is proportional to the surface area A of the slab and the instantaneous absolute temperature change per unit thickness dT/dx of the material. The constant of proportionality K_T is known as the thermal conductivity of the material. The minus sign means that heat flow is in a direction of decreasing temperature.

$$dQ/dt = -K_T A dT/dx$$

7.5 The internal energy U of an ideal gas containing N molecules is proportional to the product of N and the absolute temperature T. The constant of proportionality is 3k/2 where k is Boltzmann's constant.

$$U = (3/2)NkT$$

7.6 In an idealized heated solid called a cavity radiator, the energy radiated from the cavity interior per unit area (called total cavity radiancy, R_C) is proportional to the fourth power of the absolute temperature T. The constant of proportionality σ is called the Stefan-Boltzmann constant.

$$R_C = \sigma T^4$$

8. **Temperatures and Conversions**

C^0 is the symbol for degrees Celsius, F^0 is the symbol for degrees Fahrenheit and K^0 means degrees Kelvin (Absolute).

8.1 Water freezes at 0 C^0 at standard atmospheric pressure.

8.2 Water boils at 100 C^0 at standard atmospheric pressure.

8.3 The triple point of water (existing simultaneously as a gas, liquid and solid) occurs at a temperature of 273.16 K^0 and atmospheric pressure of 611.73 Pascals (Newtons per square meter).

8.4 $C^0/100 = (F^0-32)/180$

8.5 $K^0 = C^0 + 273.16$

Electromechanical Laws

1. **Maxwell's Equations**:

1.1 The source of the electric field (**E**) is charge density ρ_q. $\nabla = (\partial/\partial x, \partial/\partial y, \partial/\partial z)$ is the normal vector operator, ($\bullet$) is the normal vector scalar product and ε_0 is a constant called the permittivity of free space. This law is also known as Gauss's law for electricity. The differential form is

$$\nabla \bullet \mathbf{E} = \rho_q/\varepsilon_0$$

The integral form is

$$\varepsilon_0 \oiint \mathbf{E} \bullet \mathbf{n} dS = q$$

where $\oiint$ means integration over the closed surface S, **n** is a unit vector normal to S enclosing the charge q.

1.1.1 Maxwell's first equation and may be used to derive Coulomb's law which states that the force between two charges is proportional to the product of the two charges and inversely proportional to the square of the distance between their charge centers. The force is in a direction on a line drawn between the two charges q_1 and q_2 denoted by the unit vector $\mathbf{r}_{u.}$ $K_C = 1/(4\pi\varepsilon_0)$ will be called Coulomb's constant.

$$\mathbf{F} = \mathbf{r}_u K_C q_1 q_2 / r^2$$

giving rise to the electrical potential energy, V between q_1 and q_2

$$V = K_C q_1 q_2 / r$$

If the charges are both positive or both negative, the force is repulsive (like charges repel one another), otherwise the force is attractive (unlike charges attract one another).

1.2 The source of the magnetic field **B** is zero. This is Maxwell's second equation. This also means that magnetic fields always exist in closed loops and magnetic monopoles do not exist. This law is also known as Gauss's law for magnetism. The differential form is

$$\nabla \bullet \mathbf{B} = 0$$

The integral form is

$$\oiint \mathbf{B} \bullet \mathbf{n} dS = 0$$

where $\oiint$ means integration over any closed surface S, **n** is a unit vector perpendicular to the surface, S.

1.3 Ampere's law is also known as Maxwell's third equation. Current density $\mathbf{J}_i$ and/or dynamic electric fields, $\partial\mathbf{E}/\partial t$ give rise to circulating magnetic fields (**B**). μ_0 is known as the permeability constant of free space. The differential form is

$$\nabla \times \mathbf{B} = \mu_0 \mathbf{J}_i + \mu_0 \varepsilon_0 \partial\mathbf{E}/\partial t$$

where $\nabla = (\partial/\partial x, \partial/\partial y, \partial/\partial z)$ is the Del vector operator and **x** is the vector cross product. The integral form is

$$(1/\mu_0)\oint \mathbf{B} \bullet \mathbf{ds} = i$$

where $\oint$ means integration over a closed line s, circulating around the electrical current, i. **ds** is an infinitesimal vector line element of s, that **B** circulates through. **B** is perpendicular to the direction of the electrical current i.

1.4 Faraday's law is also known as Maxwell's fourth equation. Dynamic magnetic fields, ($\partial\mathbf{B}/\partial t$) give rise to circulating electric fields (**E**). The differential form is

$$\nabla \mathbf{x}\, \mathbf{E} = -\partial\mathbf{B}/\partial t$$

where $\nabla = (\partial/\partial x, \partial/\partial y, \partial/\partial z)$ is the normal Del vector operator and **x** is the vector cross product.

The integral form is

$$\oint \mathbf{E}\bullet d\mathbf{s} = -\iint (\partial \mathbf{B}/\partial t)\bullet \mathbf{n}dS = -\partial\Phi/\partial t$$

where $\Phi = \iint \mathbf{B}\bullet\mathbf{n}dS$ is called the magnetic flux in which **B** penetrates the surface area S. **n** is a unit vector perpendicular to the surface area S.

2. **The Lorentz Force**:

The force **F** on a charge q moving with velocity **v** by an external electric field **E** and by an external magnetic field **B** and **x** is the normal vector cross product.

$$\mathbf{F} = q\mathbf{E} + q\mathbf{v} \times \mathbf{B}$$

3. **Electromagnetic Wave Equations:**

When there is no charges or currents, as in the vacuum of matter free space, Maxwell's equations yield a wave equation that is satisfied by both the electric field **E** as well as the magnetic field **B**.

These equations yields the precise description of induced electromagnetic fields.

3.1 $\nabla^2\mathbf{E} - \partial^2\mathbf{E}/(c^2\partial t^2) = 0$ and

3.2 $\nabla^2\mathbf{B} - \partial^2\mathbf{B}/(c^2\partial t^2) = 0$

where $\nabla^2 = \nabla \bullet \nabla = \partial^2/\partial x^2 + \partial^2/\partial y^2 + \partial^2/\partial z^2$, t is the time and c is the speed of light in vacuum. Note that if one utilizes the gradient operator, $\Box$ defined as $\Box = (\partial/\partial x, \partial/\partial y, \partial/\partial z, \partial/\partial(ict))$ then, the Dalembertian operator, $\Box^2 = \Box \bullet \Box = \partial^2/\partial x^2 + \partial^2/\partial y^2 + \partial^2/\partial z^2 - \partial^2/(c^2\partial t^2)$ makes the electromagnetic wave equations 3.1 and 3.2 simplify to

3.1.1 $\Box^2\mathbf{E} = 0$ and

3.2.1 $\Box^2\mathbf{B} = 0$

Conservation Laws

1. Conservation of energy: A system's total energy, E_T is the same both before (B) and after (A) any energy transformation.

$(E_T)_B = (E_T)_A$

2. Conservation of momentum: A system's total momentum, p_T is the same both before and after any energy transformation.

$(p_T)_B = (p_T)_A$

3. Conservation of angular momentum: A system's total angular momentum, L_T is the same both before and after any energy transformation.

$(L_T)_B = (L_T)_A$

4. Conservation of charge: A system's total charge, Q_T is the same both before and after any energy transformation.

$$(Q_T)_B = (Q_T)_A$$

5. Conservation of baryon number: A system's baryon number, N_B is the same both before and after any energy transformation. Baryons are composed of quarks. Quarks have baryon number +1/3. Antiquarks have baryon number –1/3.

$$(N_B)_B = (N_B)_A$$

6. Conservation of lepton number: A system's lepton number, N_L is the same both before and after any energy transformation.

$$(N_L)_B = (N_L)_A$$

7. For any energy system, another related energy system predicted by the simultaneous operations of time reversal, charge conjugation (signs of all

charges involved are reversed) and space reversal (mirror image or parity) is also possible. This is called CPT for short. Below, E_T is the total energy of a system and BCPT means before the CPT operation and ACPT means after the CPT operation.

$$(E_T)_{BCPT} = (E_T)_{ACPT}$$

Basic Units

position: (measured with a ruler)

meter = m

mass: (measured with a balance scale)

kilogram = kg

time: (measured with a clock)

second = s

charge: (measured with a voltmeter)

coulomb = coul

Equivalent Units

Force: Newton = nt = kg–m/s^2

Pressure: Pascal = nt/m^2

Energy: joule = nt–m

Inductance: henry = joule–m–s^2/$coul^2$

Capacitance: farad = $coul^2$/joule

Basic Physical Constants

Name	Symbol	Value
Speed of light	c	3.00×10^{8} m/s
Gravitational Constant	G	6.67×10^{-11} nt-m^2/kg^2
Avogadro's number	N_0	6.023×10^{23} /mole
Universal Gas Constant	R	8.32 joules/(mole-K^0)
(Planck's constant)/2π	$\hbar$	1.055×10^{-34} joule-s
Planck length	$L_P = (\hbar G/c^3)^{1/2}$	1.616×10^{-35} m
Planck time	$T_P = (\hbar G/c^5)^{1/2}$	5.391×10^{-44} s
Planck mass	$M_P = (\hbar c/G)^{1/2}$	2.177×10^{-8} kg
Boltzmann's constant	k	1.38×10^{-23} joules/(molecule-K^0)
Stefan-Boltzman constant	σ	5.67×10^{-8} joules/m^2/(K^0)^4
Permeability constant	μ_0	1.26×10^{-6} henry/m

Name	Symbol	Value
Permittivity constant	ε_0	$8.85 X 10^{-12}$ farads/m
electron charge	q_e	$-1.6022 X 10^{-19}$ coul
Electron rest mass	m_e	$9.11 X 10^{-31}$ kg
Proton rest mass	m_p	$1.67239 X 10^{-27}$ kg
Neutron rest mass	m_N	$1.6747 X 10^{-27}$ kg
Coulombs constant	$1/(4\pi\varepsilon_0)$	$8.99 X 10^9$ nt-m^2/$coul^2$

Basic Elementary Particles

Preliminary Particle Descriptors

1. Family Names – Particles belong to functional families having a set number of family members. For example, the gluon family has eight members and they function to provide the strong nuclear force. Individual particles have both a historical name and a symbol. For example, an electron has the symbol e^-.

2. Color – Quarks can either be red, green or blue (r,g,b). Anti-quarks can either be –red, –green or –blue (–r, –g, –b). This is similar to charge coming in two types, the minus (–) and the plus (+) type.

3. Charge – measured in units of positive electronic charge or the charge on a positron (anti–electron). The charge magnitude of a negative electron (e^-) or a positive positron (e^+) are equal. An anti–particle has the opposite charge as the particle.

4. Spin – Axial angular momentum measured in units of Planck's constant divided by 2π and denoted by $\hbar$. Quantum Spin is specified as positive, but it is understood that quantum mechanically, it can either be positive (parallel) or negative (anti–parallel) to any given direction. Fermions (matter particles) have half integral values of $\hbar$. Bosons (force field particles) have integral values of $\hbar$.

5. Helicity – Helicity is also given in terms of $\hbar$ and may be thought of as the component of the particle's spin in the direction of the particle's velocity vector. The helicity of particles moving at the velocity of light is different than the helicity of particles that do not. Particles moving at the velocity of light, c such as photons, must have zero rest mass and there is no coordinate system for which its velocity is zero. Thus, the component of a photon's spin ($\hbar$) along its velocity vector is the same as its spin orientation, either $+\hbar$ or $-\hbar$ since it cannot be observed at rest. Thus, a photon has an intrinsic helicity the same as

its intrinsic spin. On the other hand, particles with non-zero rest mass have non–intrinsic helicity dependent on the observer since their spin can be observed when they are at rest and their spin components in the direction of motion must have a quantum difference of $+\hbar$. For example, the weakons, responsible for the electroweak forces, with non-zero rest masses and spin of $\hbar$ have helicity of either $-\hbar$, 0, or $+\hbar$. A particle and its anti–particle have opposite helicity.

6. Rest Mass – Measured in either Proton rest masses (Mp) or millions of electron volts (Mev). An electron volt (1.602×10^{-19} joules) is the kinetic energy an electron gains by being propelled a distance of one meter by an electrical field of strength, one volt per meter. The equivalent energy of a proton at rest is 938 Mev. The reason rest mass can be measured in terms of energy is because of Einstein's famous equation $E_0 = m_0c^2$ which relates rest mass, m_0 to rest mass energy, E_0 by a constant, being the square of the speed of light, c^2.

7. Field Energy – Force fields are caused by corresponding field particles having integral values of $\hbar$ (called bosons). Matter particles having half integral values of $\hbar$ (called fermions) are influenced by force fields caused by their interaction with the corresponding boson. The four force fields are strong nuclear (gluons), electroweak (weakons), electromagnetic (photons) and gravitational (gravitons).

Anti–Particle Properties

All particles have an anti–particle. The anti–particle has the opposite charge of the particle. The anti–particle has the opposite helicity of the particle. The anti–particle of a non–zero rest mass particle having zero charge, and having a spin of one $\hbar$ and zero helicity is the particle itself. The antiparticle has the same mass as the particle. A particle and its anti–particle (that is not itself) annihilate one another upon contact in a burst of other energetic particles.

Matter Energy Particles

All material energy is composed of fundamental matter particles experimentally observed to exist as three energy families (UP, CHARMED, TOP) of four fermions each, in its simplest representation. Two of the fermions are light and are called leptons and two of the fermions are heavy and are called quarks. One of the leptons carries a negative electronic charge, the other has no charge.

Origin of the UP Family

The nuclei of atoms are composed of neutrons and protons. A neutron consists of two (red and blue) down quarks, ($d_R^{-1/3}$, $d_B^{-1/3}$,$u_G^{2/3}$) and one (green) up quark, . A proton consists of two (red and blue) up quarks, and one (green) down quark, ($u_R^{2/3}$, $u_B^{2/3}$, $d_G^{-1/3}$). Any other cyclic permutation of red, green or blue colored quarks in neutrons or protons is possible. The proton is stable. An isolated neutron, n is unstable and will decay into a proton, p electron, e^- and an electron anti–neutrino, $\acute{\upsilon}_e$. The net effect is that one of the down quarks of the neutron will change into an electron, anti–neutrino and an up quark. This effectively transformed the internal structure of a neutron ($d_R^{-1/3}$, $d_B^{-1/3}$, $u_G^{2/3}$) into that of a proton ($u_R^{2/3}$, $u_B^{2/3}$, $d_G^{-1/3}$). The up quark has a charge of 2/3 e^+ while the down quark has a charge of –1/3 e^+. Thus a proton has a net charge of e^+ while the neutron has a net charge of 0. The UP family making up neutrons and protons consist of four family members which are the up quark, down quark, electron and its anti–neutrino. There are two other four member families. The TOP family has the highest rest mass energy particle members. The CHARMED family has intermediate rest mass energy particle members. The UP family has the lowest rest mass energy particles. Each family maintains the same relationships between its members.

The UP Family

The UP family consists of an up quark, $u^{2/3}$, a down quark, $d^{-1/3}$, electron, e^-, with its electron anti–neutrino, $\acute{\upsilon}_e$. The quarks can either be red, blue or green. The up quark has a charge of 2/3 e^+ while the down quark has a charge of –1/3 e^+. The electron has a rest mass energy of .511 Mev. These particles have the lowest rest mass energy and represent the ground state rest mass energy of the matter families. All UP fermion family members have a spin of ½ħ and a helicity of plus or minus ½ħ.

The CHARMED Family

The CHARMED family consists of a charmed quark, $c^{2/3}$, a strange quark, $s^{-1/3}$, muon, μ^- with its muon anti–neutrino, $\acute{\upsilon}_\mu$. The quarks can either be red, blue or green. The charmed quark has a charge of 2/3 e^+ while the strange quark has a charge of –1/3 e^+. The muon has a rest mass energy of 105.66 Mev. These particles have intermediate energy and represent a higher rest mass energy state than the UP family. All CHARMED fermion family members have a spin of ½ħ and a helicity of plus or minus ½ħ.

The TOP Family

The TOP family consists of a top quark, $t^{2/3}$, bottom quark, $b^{-1/3}$, tauon, τ^{-} with its tauon anti–neutrino, $\acute{\upsilon}_{\tau}$. The quarks can either be red, blue or green. The top quark has a charge of 2/3 e^{+} while the bottom quark has a charge of –1/3 e^{+}. The tauon has a rest mass of 1784.2 Mev. These particles have the highest rest mass energy state and represent a higher energy state than that of the CHARMED family. All TOP fermion family members have a spin of ½$\hbar$ and a helicity of plus or minus ½$\hbar$.

Field Energy Particles

Gluon Family

Gluons (g_1 – g_8) are responsible for the strong force field between the three colored (red, green and blue) quarks making up protons and neutrons, of which all nuclei are composed. There are eight different gluons. Gluons carry color combinations (r, g, b, –r, –g, –b) and compose the gluon field holding quark trios together in protons and neutrons. Gluons have a spin of $\hbar$. Gluons have zero rest mass and therefore move at c, the velocity of light. Thus, gluons have helicity of either plus or minus $\hbar$.

Photon Family

Photons (γ) are responsible for the electromagnetic forces which act between charges. Photons have no color and no charge. Photons have a spin of $\hbar$. Photons have no rest mass and move at the velocity of light. Thus, photons have helicity of either plus or minus one $\hbar$. The positive helicity photon is the anti–photon of the negative helicity photon. While in flight, photons have mass, energy and momentum.

The Weakon Family

Weakons give rise to the electroweak force field responsible for radioactive decay. Recall that a neutron is composed of two down quarks and one up quark. The decay of an isolated neutron is an example of radioactive beta (electron) decay in which one of the down quarks in a neutron decays into a weakon (the omega minus) which then decays into an up quark, electron and anti–neutrino. The net effect is that a neutron decays into a proton, electron and anti–neutrino. There are three different weakons, the omega minus (Ω^-), omega zero or zeta (Z^0) and the omega plus (Ω^+). These weakons have no color and carry charges of e^-, 0, e^+ respectively. Weakons have a spin of $\hbar$ and each can have helicity of $-\hbar$, 0 or $+\hbar$. Weakons have rest masses of 85 Mp, 260 Mp and 85 Mp respectively. Anti–weakons have opposite charges and helicities as the corresponding weakons.

The Meson Families

Mesons give rise to the forces between baryons (quark trios). Mesons are not elemental but are composed of quark anti–quark pairs (combos taken from any of the three families of quarks) and are mentioned here for completeness. Obviously, there are many families of mesons, and the pi meson family (pions) are responsible for forces between nucleons (either neutrons or protons). Pions will be presented next as an example.

Pi Meson (Pions) Example

The pi mesons are the example of the family which will be discussed. Recall that neutrons and protons are composed of down quarks and up quarks. The pi minus (π^-) is composed of a down quark with a charge of –1/3 e^+ and an anti–up quark with a charge of –2/3 e^+ for a total charge of e^-. The pi zero (π^0) is a mixture of an up quark and an anti-up quark, or a down quark and an anti–down quark. The pi plus (π^+) is composed of an up quark with a charge of +2/3 e^+ and an anti–down quark with a charge of +1/3 e^+ for a total charge of e^+. These pions have no color and carry charges of e^-, 0, e^+ respectively. Pions have a spin of 0 and each has helicity of 0. The charged pions have rest masses of 139.57 Mev, while the pi zero has a rest mass of 134.96 Mev. The anti–pi minus is the pi plus. The anti–pi plus is the pi minus. The anti pi zero is the pi zero itself.

Graviton Family

Gravitons (G_- and G_+) are responsible for the gravitational force fields which act between masses. Gravitons have no color and no charge.

The G_- graviton has a spin of $2\hbar$ and a zero rest mass. It moves at the velocity of light and thus, its helicity is $-2\hbar$ or $+2\hbar$. It is assumed to have negative mass in flight while being exchanged between any two positive masses or any two negative anti–masses. This is because the gravitational potential energy between two positive masses or two negative masses is negative.

Because of a new scientific theory called "Nature of the First Cause", in which positive matter is gravitationally repelled by negative anti–matter, the G_+ graviton is postulated to exist. It also has a spin of $2\hbar$. It is assumed to have a zero rest mass and moves with the velocity of light and has positive mass in its flight between negative antimatter and positive matter. Thus, the G_+

graviton also has helicity of $-2\hbar$, or $2\hbar$. The G_+ gravitons fill up all space and are responsible for the force of repulsion between negative anti–matter and positive matter. By the "First Cause" theory, it makes up the repulsive gravitational field which is responsible for the accelerated expansion of distant positive matter in the universe (galaxies not in the local group).

The Higgs Family

There are two Higgs bosons (H_L and H_H) called the light Higgs boson, H_L of the unified electroweak theory and the heavy Higgs boson, H_H of the grand unified theory. The heavy Higgs boson, makes up the Higgs field and permeates all space. This field is responsible for assigning masses to all the fundamental particles. The light Higgs boson is responsible for assigning the masses to the weakons. Both Higgs bosons have a spin of zero (0), and thus they both have a helicity of zero. Both Higgs bosons have non–zero rest masses with the light Higgs rest mass at roughly 10^5 Mev and the heavy Higgs rest mass of about 10^{17} Mev.

Complete Set of Particles

All matter particles which have been discovered are combinations of the above elementary matter particles. All the known force fields consists of varying energy and intensity of the above force field particles.

The Hadrons (consisting of quarks) which are matter particles that have been discovered now number over two hundred which exceed the number of known elements.

References

Al-Khalili, Jim, *Quantum, A Guide for the Perplexed*, United Kingdom, Weidenfeld & Nicolson, 2003

Ames, Joseph Sweetman & Murnaghan, Francis D., *Theoretical Mechanics An Introduction to Mathematical Physics*, New York, Dover Publications, Inc., 1957

Atkins, K. R., *Physics*, New York, John Wiley & Sons, Inc., 1965

Bergmann, Peter Gabriel, *Introduction to the Theory Of Relativity*, New York, Dover Publications, Inc., 1976

Blass, Gerhard A., *Theoretical Physics*, New York, Appleton-Century-Crofts, 1962

Born, Max, *Einstein's Theory of Relativity*, New York, Dover Publications, Inc., 1962

Breithaupt, Jim, *Cosmology*, Blacklick, OH, McGraw-Hill, 1999

Davies, Paul, *The New Physics*, New York, Cambridge University Press, 1996

De Broglie, Louis, *matter and light*, New York, Dover Publications, Inc., 1939

Einstein, Albert, *Builders of the Universe*, Los Angeles, CA, U. S. Library Association, Inc., 1932

Einstein, Albert, & Lorentz, H. & A., Minkowski, H., & Weyl, H., *The Principle of Relativity*, New York, Dover Publications, Inc., 1952

Einstein, Albert, *Relativity The Special and General Theory*, New York, Crown Publishers, Inc., 1961

Feynman, Richard P., *QED The Strange Theory of Light and Matter*, Princeton, New Jersey, Princeton University Press, 1988

Feynman, Richard P., *Six Not So Easy Pieces*, New York, Basic Books, 1997

Frankel, Theodore, *Gravitational Curvature An Introduction to Einstein's Theory*, San Francisco, W. H. Freeman and Company, 1979

Goldstein, Herbert, *Classical Mechanics*, London, Addison-Wesley Publishing Company, Inc., 1950

Halliday, David & Resnick, Robert, *Physics For Students of Science and Engineering*, New York, John Wiley & Sons, Inc., 1962

Hawking, Stephen & Penrose, Roger, *The Nature of Space and Time*, New Jersey, Princeton University Press, 1996

Heisenberg, Werner Karl, *The Nature of Elementary Particles*, in Physics Today, Page 39, March 1976

Kaku, Michio, *Hyperspace*, New York, Anchor Books, 1995

Kaku, Michio, *Parallel Worlds*, New York, Anchor Books, 2006

Kaplan, Irving, *Nuclear Physics*, Reading, Massachusetts, Addison-Wesley Publishing Company, Inc., 1962

Korn, Granino A. & Theresa M., *Mathematical Handbook For Scientists And Engineers*, New York, McGraw–Hill Book Company, 1968

Lindsay, Robert Bruce, *Physical Mechanics*, Princeton, New Jersey, D. Van Nostrand Company, Inc., 1961

McMahon, David, *quantum mechanics demystified*, New York, McGraw Hill, 2005

Messiah, Albert, *Quantum Mechanics*, New York, Dover Publications, Inc., 1999

Park, David, *Introduction to the Quantum Theory*, New York, McGraw-Hill Book Company, 1964

Penrose, Roger, *The Road To Reality, A Complete Guide to the Laws of the Universe*, New York, Vintage Books, 2004

Powell, John L. & Crasemann, Bernd, *Quantum Mechanics*, Reading, Massachusetts, Addison-Wesley Publishing Company, Inc., 1961

Ridpath, Ian, *The Illustrated Encyclopedia of the Universe*, New York, Watson-Guptil Publications, 2001

Sears, Francis W., & Zemansky, Mark W. & Young, Hugh D., *College Physics*, Menlo Park,

California, Addison-Wesley Publishing Company, 1986

Segre, Emilio, *Nuclei and Particles*, New York, W. A. Benjamin, Inc., 1964

Shortley, George & Williams, Dudley, *Elements of Physics For Students of Science and Engineering*, Englewood Cliffs, New Jersey, Prentice-Hall, Inc., 1965

Symon, Keith R., *Mechanics*, Reading, Massachusetts, Addison-Wesley Publishing Co., Inc., 1961

Van Heuvelen, Alan, *Physics, A General Introduction*, Boston, Little, Brown and Company, 1982

Weinberg, Steve, *Dreams of a Final Theory: The Search for the Fundamental Laws of Nature*, New York, Pantheon Books, 1992

Weyl, Hermann, *Symmetry*, Princeton University Press, 1952

Young, Hugh D., *Statistical Treatment of Experimental Data*, McGraw-Hill Co., Inc., 1962

INDEX

A

B

C

D

E

F

G

H

I

K

L

M

N

O

P

S

T

U

V

W

Z

17729787R00107

Made in the USA
Middletown, DE
06 February 2015